Books should be returned or renewed by the last
date above. Renew by phone **03000 41 31 31** or
online *www.kent.gov.uk/libs*

Libraries Registration & Archives

The Secret Network of Nature

By Peter Wohlleben

The Mysteries of Nature Trilogy

The Hidden Life of Trees
The Inner Life of Animals
The Secret Network of Nature

The Secret Network of Nature

The Delicate Balance of All Living Things

PETER WOHLLEBEN

Translated from the German by Jane Billinghurst

THE BODLEY HEAD
LONDON

1 3 5 7 9 10 8 6 4 2

The Bodley Head, an imprint of Vintage,
20 Vauxhall Bridge Road,
London SW1V 2SA

The Bodley Head is part of the Penguin Random House group of companies
whose addresses can be found at global.penguinrandomhouse.com.

Copyright © Ludwig Verlag, Munich, part of the
Random House GmbH publishing group, 2017

First published in Germany as *Das Geheime Netzwerk der Natur* in 2017

First published in Great Britain by The Bodley Head in 2018

www.penguin.co.uk/vintage

A CIP catalogue record for this book is available from the British Library

ISBN 9781847925244 (hardback)
ISBN 9781847925251 (trade paperback)

Typeset in 11.28/14.6 pt Bell MT
by Integra Software Services Pvt. Ltd, Pondicherry

Printed and bound in Great Britain by Clays Ltd, Elcograf S.p.A.

Penguin Random House is committed to a sustainable future for our
business, our readers and our planet. This book is made from Forest
Stewardship Council® certified paper.

Contents

Introduction

NATURE IS LIKE A GIANT clockwork mechanism. Everything is neatly arranged and interconnected. Everything has its place and its function. Take the wolf, for example. Under the order Carnivora, there is the suborder Caniformia, which includes the family Canidae and the subfamily Caninae, which includes the genus *Canis*, and within that genus we finally get to the species wolf. Phew. As predators, wolves keep down the number of plant eaters so that deer populations, for example, do not multiply too rapidly. All animals and plants are part of a delicate equilibrium, and every entity has its purpose and role in its ecosystem. For us humans this way of organising life supposedly gives us a clear view of the world, and thus a sense of security. As erstwhile plains dwellers, our most important sense is sight, and so our species relies on viewing things clearly. But do we really have a clear view of what is going on?

The wolves remind me of an episode from my childhood. I was about five years old and on holiday visiting my grandparents in Würzburg when my grandfather gave me an old clock. The first thing I did was take it apart: I just couldn't wait to find out how it worked. But even though I was convinced I knew how to put it back together in working order, I discovered I couldn't. I was just a kid, after all. So after I'd finished reassembling it, there were still a few cogs left over, and my grandfather was not best pleased.

In the wild, wolves are those cogs. If we eradicate them, not only do we get rid of the enemies of sheep and cattle ranchers, but the finely tuned mechanism of nature begins to run differently – so differently that rivers may change course and local bird species die out.

Things can also go awry when a new species is added to a habitat: the introduction of a non-native fish, for example, can lead to a massive reduction in the local elk population. Because of a *fish*? The earth's ecosystems, it seems, are just a bit too complex for us to compartmentalise them and draw up simple rules of cause and effect. Even conservation measures can have unexpected results. Would you expect, for example, that the recovery of crane populations in Europe affects the production of Iberian ham?

It's high time, therefore, that we took a good look at the interconnections between species both large and small. In so doing, we get the chance to meet a range of rather odd creatures, such as red-headed flies that take wing only at night in winter on the lookout for old bones, or beetles that seek out cavities in rotting trees, where they dine on the feathery remains of pigeons and owls (but only when they're mixed together). The more light you shed on relationships between species, the more fascinating facts you discover.

But nature is much more complex than a clock. In nature, not only does one cog connect with another: *everything* is connected in a network so intricate that we will probably never grasp it in its entirety. And that is a good thing, because it means that plants and animals will continue to amaze us. It's important for us to realise that even small interventions in nature can have huge consequences, and we'd better keep our hands off anything that we have no pressing reason to touch.

So for you to get a clearer picture of this intricate network, I'd like to show you some examples. Let's be amazed together.

1

Of Wolves, Bears and Fish

WOLVES ARE A WONDERFUL example of how complex the
connections in nature can be. For amazingly enough,
these predators are able to reshape riverbanks and change
the course of rivers.

This is what happened in Yellowstone, the very first
national park in the United States. In the nineteenth
century, people began to systematically eradicate wolves
in the park, primarily in response to pressure from
ranchers in the surrounding area, who were worried
about their grazing livestock. The last pack was wiped
out in 1926. Individual wolves continued to be spotted
occasionally until the 1930s, when they, too, were elimi-
nated. Other animals living in the park weren't harmed;
on the contrary, some were actually looked after. In harsh
winters, for example, rangers even went as far as feeding
the elk.

It wasn't long before the consequences became clear.
No sooner was the pressure from predators lifted than
elk populations began to increase steadily, and large areas
of the park were stripped bare by the voracious animals.
Riverbanks were particularly hard hit. The juicy grass

by the river disappeared, along with all the saplings growing there. Now this desolate landscape didn't provide enough food even for birds, and the number of species declined drastically. Beavers were among the losers, because they depend not only on water but also on the trees that grow by the river – willows and poplars are some of their favourite foods. They cut them down so they can get at the trees' nutrient-rich new growth, which they devour with relish. Because all the young deciduous trees alongside the water were ending up in the stomachs of hungry elk, the beavers had nothing to gnaw on, and they disappeared.

Riverbanks became wastelands, and without any vegetation to protect the ground, seasonal flooding washed away ever-increasing quantities of soil. Erosion advanced rapidly. As a result, the rivers began to meander more and follow increasingly winding paths through the landscape. And the less protection from erosion for the underlying layers of soil, the more pronounced that serpentine tendency, especially in a flat landscape.

This sorry state of affairs continued for decades or, to be precise, until 1995. This was the year that wolves caught in Canada were released in Yellowstone to restore the park's ecological balance. Scientists call what happened in the years that followed, and continues to this day, a trophic cascade: a change in the entire ecosystem all the way down the food chain. But with the wolf now at the very top, what was let loose could perhaps better be described as a trophic avalanche.

The wolves did what we all do when we're hungry: they looked for something to eat. What they found in the park were large numbers of easy-to-catch elk. It seems obvious where this story's heading: the wolves ate the elk and elk numbers declined drastically, which gave little trees a chance to grow again. Does that mean that the key to solving the problem of disappearing trees is to replace elk with wolves? Thankfully, it's not nature's way to simply swap out one animal for another, and here's why. The fewer elk there are, the longer it takes the wolves to find them, and below a certain residual number, it's no longer worth their while to hunt elk, which means that the wolves must either leave the park or starve.

In Yellowstone, however, in addition to declining elk numbers, there was something else going on. Thanks to the presence of wolves, the elk's behaviour was changing, and that change was triggered by fear. Elk began avoiding open areas along the riverbanks and retreating to places that offered more cover. True, they did come to the water from time to time, but they no longer stayed long; and when they were there, they were constantly scanning the landscape, worried they might spot one of the grey-coated hunters. Such constant surveillance left them little time to get their heads down among the willow and poplar saplings now growing abundantly along the riverbanks. Both trees are so-called pioneer species, and they grow faster than most trees: it's not unusual for them to grow one metre in a single season.

Within a few years, the riverbanks became stable again, slowing the flow of rivers, which in turn carried off less soil. The meandering stopped, though the serpentine curves the rivers had already carved into the landscape remained. Most importantly, the beavers' food sources returned, and the industrious little creatures began to build dams, which slowed the flow of water even more. Increasing numbers of ponds formed, creating small paradises for amphibians. This blossoming diversity also saw the number of bird species increase substantially. (You can find an impressive video about this on the home page for Yellowstone National Park.)[1]

There are some, however, who question the cause and effect here. At the same time the wolves returned, a drought that had lasted many years ended, and the heavier rainfall was better for the trees – willows and poplars in particular love moist soil. But this explanation for why the trees grew better ignores the beavers. In the places where these buck-toothed engineers live, variations in precipitation have little effect on tree growth, especially close to riverbanks. Beaver dams hold back the water, but at the same time help saturate the banks and make it easier for trees to find water even when it hasn't rained for months. And it's precisely this process that was set in motion once again by the wolves: fewer elk by the riverbanks means more willows and poplars means more beavers. Surely that explains it?

Unfortunately, I have to disappoint you, and now things get even more complicated. Some researchers believe it's simply the number of elk that is the problem and not the way they behave. In this scenario, there are fewer elk in the park overall since the reintroduction of wolves because so many of them have been eaten – and that's why you don't see so many of them near the rivers.

Now you're completely confused? No wonder. I have to admit that for a while there even I felt like that five-year-old who took the clock to bits all over again. In the case of Yellowstone, however, there's no doubt that because human intervention has been dialled back, the clock is slowly beginning to tick again. And for scientists to admit they don't yet understand this process in every detail is actually encouraging. The more we acknowledge that even the smallest disturbance can lead to unpredictable changes, the stronger the arguments for protecting larger areas of the environment.

The reintroduction of wolves has done more than just help the trees and creatures along the riverbanks. Other predators have benefited too. In the decades when the elk population was exploding things were not looking so good for grizzlies. In the autumn, grizzly bears depend on berries. They feed greedily on these tiny sugar- and carb-filled power snacks in order to pile on the pounds before winter comes. At some point, however, the small shrubs that had seemed to offer an inexhaustible supply of berries were no longer providing enough, or, rather, the bushes were being plundered by someone else – for elk also love this calorie-rich fruit. Now wolves are hunting these large plant eaters again, there are more berries left for the bears. And since the arrival of the wolves, the bears have been in much better health.[2]

I began the wolves' story by saying that they were eradicated in response to pressure from ranchers. But when the wolves disappeared, the ranchers did not. They still graze their livestock on range land that runs right up to the national park boundary. The attitude of many of them hasn't changed over the past decades either, so it's no surprise that wolves are shot the moment they leave the park. In recent years their number has dropped sharply, from a high point of 174 individuals in 2003 to about 100 in 2016.

One reason is improvements in technology. Many Yellowstone wolves now wear radio collars, to help researchers locate the packs and find out how they move around the park, or when they leave it altogether. These radio signals are intercepted illegally, I was told by Elli Radinger, who has spent years observing wolves in Yellowstone, to enable people to shoot them the moment they leave the protection of the park. There's no more effective way of hunting down wolves. Hunters in Germany, it seems, have got in on the act too: in 2016, a young wolf wearing a radio collar was killed on the Lübtheener Heide nature reserve in Mecklenburg-Vorpommern.[3] It's ironic that the same tool that helps scientists get a better understanding of wolf movements so they can protect them is also being used by hunters to track them down and kill them.

Despite such bad news, wolves remain goodwill ambassadors for the whole notion of protecting the environment. That the return of wild animals this large to a region as densely populated as central Europe has been not only accepted by people but positively embraced is almost unbelievable. That is a blessing for all nature lovers, and even more for nature itself.

Many parts of central Europe are in a similar situation to Yellowstone before the wolves were reintroduced. Huge populations of deer and wild boar roam around unregulated by Wolf & Co. And, like the elk in America, they're consuming large quantities of food. Harsh winters barely affect the populations, and even weak animals survive and procreate happily. But here the food doesn't come courtesy of forest rangers: it's the hunters who cart massive amounts of corn, beets and hay into the forests to ensure the open-air warehouse is always fully stocked with game they can hunt.

Rangers may not be trucking in food, but the forestry service is playing its part. European forests are heavily exploited, and tree felling on such a large scale allows so much light through to the forest floor that grasses and other plants that spring up supplement the feeding programme, encouraging still further population growth. Today, there are fifty times more roe deer in German forests than there used to be in ancient times; and red deer, originally animals of the plains, now seek the safety of the trees too because humans are inhabiting their ancestral ranges. With all these large mammals eating most of the saplings, in most places the forest can no longer regenerate naturally.

Bad for the forest, but good for the wolf. The returning wolves are finding a pantry stuffed full of tasty treats. And their prey have completely forgotten how to react appropriately to the new threat – for more than a hundred years, human hunters have been their only enemy. But compared to most forest dwellers, people are slow and can't hear very well. The keenest of their senses is sight – during the day, at least – which is why countless generations of large plant eaters have learned to hide in the undergrowth by day and come out only at night. So well has this tactic worked that in relation to its size Germany – hard to believe this – is home to more large wild mammals than almost any other country on earth. And now along comes the wolf, with an altogether different hunting technique.

The first thing the wolves do is snap up 'wimpy' prey like mouflon sheep. Scientists argue over whether this sheep is truly wild or simply a domesticated sheep gone feral. Released centuries ago on islands in the Mediterranean, mouflon sheep have now advanced into central Europe, thanks to their impressively curved horns, curling almost into a circle, a desirable hunting trophy that looks good next to the deer antlers over the fireplace. Mouflon sheep are still being released into the wild today, even though it's illegal (usually the claim is that the fence around their enclosure has been 'breached').

Not only are mouflon sheep not native to central Europe, but recent developments support the contention that they are descended from domesticated stock: for wherever wolves appear, the sheep disappear – into the wolves' stomachs. The sheep, it seems, have forgotten how to take evasive action. Further damning evidence is that they are adapted for life in the mountains. Mouflon sheep are skilled climbers, and evade their pursuers by taking refuge on steep cliffs where flatlanders like wolves don't stand a chance. In forests on level ground, however, the sheep can't exploit this advantage and, when it comes to speed, they are hopelessly outmatched by the wolves. And so the mouflon sheep in Germany are caught unprepared and, as the incomers vanish, the natural order is re-established.

If mouflon sheep are such easy prey, then what about domesticated sheep or goats or calves? After all, most of them just stand around inside flimsy fences that might prevent them from running away but are easy for wolves to crawl under or jump over. But rather than turning to sensational tabloid headlines, which just love to report alleged wolf attacks on our livestock (more on that later), we would do better to peek over the shoulders of scientists. They have been analysing scat from the east German Lusatian wolves, one of the densest and longest established populations in Germany.

After collecting thousands of samples, researchers at the Senckenberg Museum of Natural History in Görlitz came to the following conclusion, which will probably surprise you. Roe deer, not sheep or goats, make up the lion's share of the wolves' diet – over 50 per cent. Red deer and wild boar account for another 40 per cent more. And no, domesticated animals still don't come next! That honour goes to hares and other small mammals, at around 4 per cent. The fallow deer, which weighs in at 2 per cent, is, much like the mouflon, an exotic species that has been released into the wild so people can hunt it, and wolves enjoy sending them to the great hunting-ground in the sky. Only now in the palette of prey do we come to a few scattered livestock animals, and they account for a meagre 0.75 per cent.[4]

But if you leaf through the pages of the tabloid press you get a completely different impression. Here, reports of wolf attacks on livestock predominate, and each and every one is worth a headline. Even before the results of genetic analysis are in to ascertain whether the culprit really is a wolf and not maybe a feral dog, the news is spread far and wide. If it turns out that the attacker was not a wolf after all, it merits only a tiny, inconspicuous correction, and the public is left with the impression that every single goat and sheep is in mortal danger.

But things don't need to be like this. It's relatively easy to keep wolves away from treasured livestock. In most cases all it takes is a simple electric fence, which lots of farmers use anyway to keep their animals from straying: the kind of coarse-mesh fence with thin strands of metal woven into it that are electrified when the fence is plugged into a charger.

We've enclosed our goat pasture at home with a fence like this, and many's the time I've forgotten to turn off the current before entering the field. Ow! The shock makes you feel like you've been hit in the back with a plank. For days afterwards I'll check again and again to make sure there's no juice running through the wire before I touch it! Imagine how much worse it feels for wolves, who'll touch the electric fence with their nose or ears. From then on, they prefer to help themselves to a side of venison or wild boar rather than face such pain again. The important thing is to make sure the fence is tall enough and in good working order. Some experts think 90 centimetres are sufficient, but my wife and I prefer to play safe, and opted for a fence that's 120 centimetres high.

Furthermore, Elli Radinger, the wolf expert I consulted, told me that if older members are shot and killed, packs can actually change their preferred range of prey. Instead of hunting wild boar, roe deer or red deer, they might switch to sheep and other domesticated animals. So those who hate wolves and want to keep them from attacking livestock would do better to leave their guns safely stashed in their cabinets.

There is yet another good thing about wolves: they bring a certain intensity to every forest experience. I still remember how happy and excited I was when I found wolf tracks one day. No, not here in Hümmel, where I live with my family, but on an out-of-the-way forest path in central Sweden. These tracks alone were enough to turn my walk into an adventure, and made the forest itself seem a little bit wilder. It's a sentiment I share with many other people for whom the wolf restores the forest's wild soul. What it demonstrates is that, even in heavily populated parts of the world, it's possible to allow the return of sizeable animals that disappeared long ago. In contrast to the situation in Yellowstone, though, wolves in Germany are returning of their own accord, moving in from Poland and slowly extending their range across the country.

Does this mean you need to be afraid every time you go for a walk in the woods? Again, newspaper reports of troublesome wolves are stacking up. Not that they've actually hurt anyone: their mere presence close to villages or, worse still, primary schools is enough to make some people's blood run cold. It goes without saying that wolves are wild animals and it's not a good idea to pet or cuddle them, but as long as we don't try to make them get used to our presence, the risks are small.

Unfortunately, there will always be some people who are tempted to feed them. That's probably what happened with the wolves Kurti and Pumpak, which kept approaching villages close to Münster and in Lausatia respectively. Both wolves ended up being shot, even though nothing they did was a threat. The fault lay not with the animals but with the people who had been offering them food.

But let's consider the situation from a different perspective. How dangerous would it be if, sometime in the future, we had not a few hundred but a few thousand wolves loping through our forests? Strictly speaking, we've been living with a far more worrying situation for a long time already, because not only our countryside but our cities too are already overrun by wolves. I'm talking about our dogs, which differ from their wild ancestors in just one important respect: they are no longer afraid of us. If you meet a wolf, it's probably just curious and will disappear when it finds out what it's dealing with. Wolves just don't think of us as prey.

It should come as no surprise, therefore, when I tell you that the real problem is dogs. If I had the choice between meeting a stray German shepherd and a wolf, I'd go for the wild animal any time. According to Olaf Tschimpke, president of the German conservation organisation NABU, 10,000 incidents of dogs biting people are reported every year, some so severe that the victims die of their injuries.[5] Imagine if just a fraction of those bites were inflicted by wolves: someone would be demanding that all wolves be shot.

Right now, however, it's not wolves that are dominating the headlines but wild boar. Even in the centre of Berlin, these pigs are calmly ploughing up lawns while people stand just a few yards away clapping and shouting to try and scare them off. Tulip beds laid waste, vineyards and cornfields stripped bare – in many places boar have become a real nuisance and are causing expensive damage. For years now, the wild boar population has been rising sharply. In Germany they have no natural enemies – or to be more precise, they *had* no natural enemies. Now that the wolf has reappeared, however, they once again have an adversary to be reckoned with.

Years ago, while walking in a former opencast coal mine one day, I came across the evidence of wolves: a pile of white bones and thick black hair that was clearly the remains of a wild boar. I realised there and then just how hard a wolf's life must be – how it has to risk its life every time it needs to eat. Because that heap of boar's remains made me think back to the hunts I used to take part in as a beater. On one occasion the dogs came across wild boar in the undergrowth and immediately gave chase. That evening, only three of the five dogs returned. The other two had probably lost their lives fighting the boar. Many handlers that deploy hunting dogs insist that the local vet is informed and available, but at the end of the day, after the work is done, some quickly sew up their dogs' wounds themselves – wounds inflicted by the wild boar's sharp tusks.

For wolves, however, even minor wounds can be life-threatening, because a wolf that has to hunt with any kind of disadvantage is likely to starve. It's truly admirable how over their ten-year life span they overcome day in, day out all the dangers they come up against.

Before we leave the subject of wolves, I want to return to Yellowstone to describe one more change – though I could pick anywhere in the world that's covered with vegetation and contains an abundance of animals. In theory central Europe would also fit the bill, but in reality there has to be a large enough area – many thousands of square miles, no less – where people are not intervening in the landscape in some way. Sadly, in Europe such areas no longer exist.

But doesn't central Europe have plenty of national parks? Yes, but by nature's standards these reserves are tiny. In most of them there's not enough room for even a single wolf pack, which means there's no opportunity to observe natural processes at work. And even in national parks, unfortunately, huge interventions can happen. In some national parks in Germany, for example, some of largest clear-felling of forests – considerably more than is normal in commercial forests – is being carried out. The powers-that-be call these 'development zones'. Even if the felling is being undertaken with the best of intentions, it still means that people are constantly interfering with natural processes.

The only way to see how nature might surprise us is to simply sit back and let things take their course, perhaps while at the same time carefully helping eradicated species reclaim territory and encouraging alien species to relinquish it. As this isn't happening in densely populated places like Germany, we must look for such success stories in other parts of the world – which brings us back to Yellowstone.

This time, fish are the centre of the story, or to be precise, lake trout. Lake trout are native to both Canada and the US (for example, in the Great Lakes), where their populations have declined significantly. As they are now endangered, there are costly hatchery programmes to help sustain wild populations. However, these lake dwellers are not endangered everywhere: in some places they have actually become a threat. No one knows who was responsible – whether anglers who wanted to increase the range of species available for them to fish, or people who just had a rather hazy grasp of conservation – but in the 1980s, lake trout suddenly appeared in Yellowstone Lake.

This wouldn't be a problem, were it not that this ecosystem was already home to one of the lake trout's smaller relatives: the cut-throat trout. The name 'cut-throat' comes from the fish's blood-red lower jaw, but it now also describes the struggle they are engaged in. For the newcomers are out-competing the original inhabitants and crowding them out, and the struggle is affecting more than just the cut-throat. Amazingly enough, the elk in the park are also suffering.

So what could possibly connect elk – which are strict vegetarians – and fish? Once again, the key to the puzzle is an intermediary, and in this case, the grizzly bear. Grizzly bears love to eat cut-throat trout, but since the introduction of their lake trout cousins they have become scarce. Cut-throat spawn in small streams, where they are easy to catch. The invasive trout behave quite differently. They shun crystal-clear tributaries, and instead lay their eggs on the lake bed, which means that the exhausted parent fish end up well beyond the reach of grizzly bears. So the bruins have to find other prey to fill their rumbling stomachs. This prey is waiting for them on land, and is trickier to hunt. The bears are now targeting elk calves, more and more of which meet their end after a blow from a grizzly bear paw. This is now happening so often that the elk population is declining noticeably.[6]

But surely, this is something to be celebrated? Weren't we just welcoming the return of the wolves for doing just this – reducing exploding elk populations? Aren't the bears, in their own way, doing the same thing? Once again, the situation isn't quite that simple. Whereas the wolves hunt older animals, the bears target young ones, which drastically alters the age distribution of the herds. To put it another way: because of the bears, the elk population is ageing, which speeds up the rate of population decline. Good for the trees, bad for the elk.

It's another example of how multifaceted ecosystems are, and how changes never affect just one species. Is it indeed possible that the wolf doesn't have the greatest influence in Yellowstone after all, and that the prize should go to the trout-bear duo instead? That enormous clock, it seems, contains more cogs than we ever suspected.

Speaking of fish, the way they interlock with the cog mechanism at work in the forest is so important that they deserve a chapter all to themselves.

2

Salmon in the Trees

_____

THE RELATIONSHIP BETWEEN TREES and fish shows just how complicated ecosystems can be. Tree growth can be almost completely dependent on these flashes of silver, especially in places where the soil is low in nutrients. Fish and rivers, it turns out, play an important role in nutrient distribution.

Let's take a look at salmon. Young salmon swim out into the ocean, where they remain for two to four years. They hunt and hang out, but mostly what they are doing is getting bigger and fatter. On the north-west coast of North America there are a number of different species of salmon, of which the king salmon (also known as Chinook) is the largest. After its youthful years at sea, a full-grown king can be up to 1.5 metres long and weigh up to 30 kilograms. After scouring the vastness of the ocean in search of food, not only has it built up a lot of muscle, but it has also stored a lot of fat, which it will need to survive its strenuous journey back to the river where it was born.

Salmon battle their way against the current to-wards the headwaters of these rivers, sometimes for many hundreds of miles and up numerous waterfalls. They carry considerable quantities of nitrogen and phosphorus in their bodies, but these nutrients are of no significance to the salmon themselves. The only reason they are toiling their way upstream is so they can spawn, in the one and only frenzy of passion they will ever experience, and then finally breathe their last.

Over the course of their journey, the salmon's silvery skin loses its metallic sheen and takes on a reddish hue. The fish are no longer eating, and as they deplete their stores of fat they are steadily losing weight. Using the last of their strength, they mate in their natal streams, and then, exhausted, they die. For the forest all around and its inhabitants, the salmon runs mean it's time to get out and haul in the catch. Lining the riverbanks are hungry hunters – bears. And along the Pacific coast of North America, this means black bears and brown bears. The fish they catch from the rapids as the salmon fight their way upstream help them put on a thick layer of fat for the winter.

Depending on location and timing, the salmon have already lost some weight by the time they're caught. At first the bears eat most of their catch, but later in the season they get choosier. They still scoop skinny salmon out of the water – fish that have used up their fat reserves and therefore contain fewer calories – but if the fish don't contain much fat, the bears don't eat much of them, and the carcasses they discard give many other animals the opportunity of a meal. Mink, foxes, birds of prey and a myriad of insects pounce on the lightly nibbled remains and drag them farther into the undergrowth.

After mealtime, some parts of the salmon (such as the bones and the head) are left lying around to fertilise the soil directly. A lot of nitrogen is also distributed through the faeces the animals expel after their feast, and overall the amount of nitrogen that ends up in the forests alongside salmon streams is enormous. According to their detailed molecular analyses, reported scientists Scott M. Gende and Thomas P. Quinn, up to 70 per cent of the nitrogen in vegetation growing alongside these streams comes from the ocean – in other words, from salmon. Their data also show that nitrogen from salmon speeds up the growth of trees to such an extent that Sitka spruce in these areas grow up to three times faster than they would have without the fish fertiliser.[1] In some trees, more than 80 per cent of the nitrogen they contain can be traced back to fish. How can we know this so precisely? The key is the isotope nitrogen-15, which in the Pacific Northwest is found almost exclusively in the ocean – or in fish. Finding evidence of these molecules in plants therefore allows researchers to make a direct connection back to the source of the nitrogen – in this case, back to the salmon.

Not all the coveted nutrients, it turns out, remain above ground. Eventually, once they've all been eaten and digested, droppings containing these nutrients are deposited on the ground, and gradually the nutrients seep down into the soil. Ready and waiting, the trees eagerly suck them up through their roots. The trees are aided by fungi that envelop their feeder roots in a fine, cottony web, which helps the trees take in the widest range of nutrients from the soil. Eventually, the trees shed their leaves or needles, and when these ancient giants die, their trunks rot back into the earth. Now an armada of organisms does a tidy job of breaking down the vegetative remains, whereupon the nutrients enter the next tree whose turn it is to draw up the released elixirs of life from the soil. Not all the nutrients remain trapped in the fine mesh of this web, however. Inevitably, some of them are washed down into the rivers and swept back out to the ocean, where innumerable tiny life-forms lie in wait.

A striking study from Japan shows how important the trees' legacy is for the oceans. Katsuhiko Matsunaga, a marine chemist at Hokkaido University, discovered that fallen leaves leach acids into streams and rivers that are then swept into the ocean. There, the acids fuel the growth of plankton, the first and most important link in the food chain. So because of the forest there are more fish? It would appear so. Matsunaga advised local fishing companies to plant trees along the coastline and river-banks: more trees would mean more leaves falling into the water and, in time, increased tree cover, leading to increased numbers of fish and oysters for these local fisheries to harvest.[2]

But let's get back to the salmon that fertilise Sitka spruce and other species living in the forests of the Pacific Northwest. It's not only the trees that are indirect beneficiaries. Dr Tom Reimchen of the University of Victoria discovered that in some insects up to 50 per cent of the nitrogen comes from fish.[3] The abundance of nutrients along salmon streams clearly shows in the increased biodiversity of animals, plants and birds, and those creatures that scavenge on salmon – foxes, birds and insects – in turn become prey for other animals in the forest.

Dr Reimchen and his team also took core samples from ancient trees. Their growth rings are like a historical archive: they reflect everything the tree has experienced over its lifetime – narrow rings for years of drought, and correspondingly wider ones for years of ample rainfall. You can also, of course, work out the level of nutrients available to the tree, and it turns out there's a direct correlation between the number of fish in earlier times and the amount of that special isotope nitrogen-15 found in wood – which is how core samples give us information about how many salmon once swam in these streams. And the answer is that their number has declined dramatically over the last hundred years; many rivers in North America today have no salmon left in them at all.

But what does all this have to do with European forests? If you're looking at how things used to be, quite a lot. European rivers were also once full of salmon, and brown bears used to roam nearby too. Unfortunately, we can't test trees from these times for nitrogen from fish, because those trees are all gone now. Since the Middle Ages, the forests were either cut down or so heavily exploited that all the ancient trees have disappeared. The average age of beeches, oaks, spruce or pines growing in Germany today is less than eighty. And eighty years ago there were neither bears nor any salmon runs to speak of, so the wood in trees in Germany doesn't contain much nitrogen-15. But what about trees from earlier times? One way to find out would be to test the beams in old timber-framed cottages for nitrogen-15, but as far as I know no one has yet done this. There's no question, however, that salmon were once plentiful in Germany, even if the evidence is largely anecdotal, such as the story that it was illegal to serve servants salmon more than three times a week.[4]

The Atlantic salmon used to be native to Europe, and thanks to the efforts of conservation organisations – particularly their efforts to clean up the waterways – it is now returning to many rivers. I grew up close to the Rhine, and my parents, I remember, didn't allow me to play in the water. Back then, chemical plants spewed out a cocktail of waste so filthy that only a few species of fish survived.

From the 1980s, regulations governing water quality were gradually put in place. Even so, in 1988 the federal minister for conservation, Klaus Töpfer, caused quite a stir when he jumped into the Rhine to swim across – three years earlier, he had placed a bet that, thanks to the new environmental policies, water quality would improve so much that the river would be fit for swimming again. When he climbed out of the brown water, the German news magazine *Der Spiegel* reported rather derisively, the minister's eyes were bloodshot. Apparently the river wasn't quite as clean as he'd hoped.[5]

Luckily, that has now changed. In 2018, the Rhine is so clean that beaches are appearing along its banks again for people to swim from. And salmon, too, are feeling at home in its waters, though they do still need help – quite a lot of help, in fact. Because adult salmon always swim back to the rivers of their youth, if all the salmon in a body of water die out, then in future salmon will almost never return there, because all mature fish were born elsewhere.

To overcome this, enterprising organisations release hundreds of thousands of young salmon into rivers where they can survive. It's not always easy to find such rivers, however, because on most of them these days dams and hydroelectric power plants impede the fish's progress. Many a turbine turns expensive hatchery-raised fish into sushi the moment they start their journey to the sea. For the return journey, there are fish ladders at dams, where the water splashes from rung to rung – i.e. from pool to pool – to imitate the rapids the fish climb by jumping up one pool at a time.

In the forest I manage, a lot of money was spent to make a small stream salmon-friendly. Barely 4 metres wide, it had been sealed off by weirs for a long time. Its name – Armuthsbach, or 'stream of poverty' – is testament to the straitened circumstances of bygone generations. Harnessing the power of the diverted water made it easier for people to grind what grain they had, and they also used the stream to fill ponds where they raised fish. Eventually, the Armuthsbach ran out of water and dried up.

Salmon are just one example of the many water-loving species, all the way down to crayfish and smaller freshwater crustaceans, that can no longer move freely when dams get in their way. And if fish and other aquatic creatures can move only downstream but not upstream, then sooner or later above the dams there will be no large life-forms in the water at all. In fact, dams are now being gradually removed so that fish can once again reach their spawning grounds. This is a huge accomplishment, and one that gives cause for hope. Indeed, adult salmon are constantly being spotted returning to the places where they were released, so that they can spawn there after spending years out at sea. Finally, after a long absence, we will be getting the first generations of truly wild salmon born in rivers to which they will return.

The salmon are returning, but unfortunately the bears are not. Bears would certainly be a problem in the big cities along the Rhine, but it's not too much of a stretch to imagine them in rural areas. However, it doesn't have to be bears that distribute fish into the landscape. What about fish-eating birds like cormorants? Cormorants were almost wiped out, but thanks to strong legal protection they are now coming back to rivers in central Europe. Since the 1990s I've been seeing them regularly on the Rhine and the Ahr. (The Ahr is a small tributary of the Rhine that has its source near the village of Hümmel where I live, and the Armuthsbach is one of the streams that flows into it.)

Cormorants are skilled divers and excellent hunters underwater. After they've eaten their fill, they doze contentedly in the tops of the trees along the riverbank. As they doze, they defecate from time to time, and their droppings naturally contain valuable nitrogen. Of course, the overall value of their excrement depends on the number of birds, and too many perching at the same time can harm the trees. This is what happened in the Saarschleife, a dramatic hairpin turn in the Saar River, where people have created a North American-style forest near the riverbank by planting Douglas firs, which come from the Pacific coast of North America. There's a whole colony of cormorants living here, and the abundant quantity of excrement they discharge is so caustic that parts of the forest canopy are dying off, much to the dismay of the local forest owner.

But that's not the main reason the birds have become so unpopular. The few salmon that battle their way upstream – products of costly reintroduction efforts – are often picked off by cormorants before they reach their spawning grounds. So what happens next? A natural nutrient cycle ensues but, as is so often the case, this cycle collides with human interests. I can understand that no one wants to stand idly by as cormorants threaten to destroy all the conservationists' hard work, but is that a good enough reason to automatically reach for a gun?

That is exactly what people did at the aforementioned Ahr, cheered on by members of the Ahr fishing syndicate, the organisation that had worked so hard on behalf of the salmon. Perhaps nature is taking a back seat here. On its home page, the syndicate (a glance indicates that membership is restricted to anglers and those who own or rent fishing waters) specifically mentions that hunting the birds, which are strictly protected under EU law, is still possible, thanks to an exemption intended to protect the fishing industry from financial loss.[6] It's a shame that this stand on cormorants taints the work of the organisation, whose efforts on behalf of salmon are thoroughly commendable.

But do forests that grow around heavily populated areas – and that includes practically all of them in central Europe – even need natural fertilisation with nitrogen? In recent decades, trees have had completely new (and completely unnatural) sources of nitrogen available to them – in fact, there's a veritable deluge of them. In contrast to the clean air in the northern United States and Canada, the air in central Europe is basically a murky soup. Perhaps not optically speaking, but definitely in terms of pollutants. Or perhaps I should say, in terms of 'nutrients'. Exhaust fumes from vehicles and manure from agriculture provide more of these than plants would like. But more of that later.

Nitrogen is naturally abundant in the air. Even as you're reading these words you're breathing large quantities in and out. Oxygen, so vitally important for us, makes up just 21 per cent of our air, compared to 78 per cent nitrogen. Strictly speaking, three-quarters of every breath we take is useless – if we could separate out the gas we don't need. That doesn't mean that nitrogen is of no use to us. On the contrary: in your body you're carrying around about 2 kilograms of it, processed into protein, amino acids, and other substances.[7]

It's pretty much the same with plants. Like us, they don't need nitrogen to breathe. What interests them are the special compounds in which it is held. These compounds are reactive, and can be broken down and converted into protein or built into plants' genetic material. In the natural world, unfortunately, these compounds tend to be rare. If a tree doesn't have the good fortune to be growing alongside a salmon stream, it has a problem. Faeces left by passing animals, or perhaps even a whole carcass rotting within reach of its roots, are therefore causes for celebration.

Lightning plays its part too, by using its energy to combine atmospheric nitrogen with oxygen to create compounds plants can break down and absorb. Some trees and plants have developed the ability to transform atmospheric nitrogen into available compounds with the help of bacteria that live in special nodules on their roots. Alders, for example, manufacture their own fertiliser this way. Most species of trees, however, can't do this, and instead rely on waste products from animals for their nitrogen needs.

Overall, it seems, nature regards these valuable nitrogen compounds as rare treats. But then we came along. Our modern internal combustion engines, in vehicles or heating systems, do the same thing as lightning. As a by-product of burning fossil fuels, they combine atmospheric nitrogen with oxygen to create compounds that travel long distances on the wind all over the world, and are then washed back down to earth when it rains. Then there's agriculture, which forces the soil to be as productive as possible by adding synthetic fertilisers that contain nitrogen. The quantity of nitrogen compounds released by human activity is significant. About 220 million tonnes rain down on the earth every year – about 27 kilograms per person in the world, and about 100 kilograms per person in industrialised countries.[8]

Maybe that doesn't sound much? Let's get back to the salmon and their beneficial effects on trees. A male chum salmon (also known as dog salmon or keta salmon) contains, on average, 130 grams of nitrogen.[9] If Europeans were to calibrate their nitrogen emissions in salmon, that would add up to around 750 fish per person per year. At 230 inhabitants per square kilometre – the population density in Germany – that would come out at 172,500 salmon per square kilometre, and such an enormous quantity of fish would quite obviously completely overload the natural cycle. Exhaust fumes and applications of liquid manure and fertiliser reach the same threshold; however, for the most part they are out of sight and therefore out of mind. They only appear on our radar when one day high levels of nitrogen compounds are detected in our drinking water.

Trees, however, have been aware of these emissions for a long time – and so have foresters. For decades, saplings have been growing markedly more quickly. This means that forests have been producing more timber, and forest production estimates now need new baselines.

Foresters' yield charts – spreadsheets that indicate how quickly and at what age different trees species grow – have already had to be adjusted upward by 30 per cent.

Is that a good sign? No. Left to their own devices, trees do not grow quickly. In undisturbed ancient forests, youngsters have to spend their first 200 years waiting patiently in their mothers' shade. As they struggle to put on a few feet, they develop wood that is incredibly dense. In modern managed forests today, seedlings grow without any parental shade to slow them down. They shoot up and form large growth rings even without a nutrient boost from added nitrogen. Consequently, their woody cells are much larger than normal, and contain much more air, which makes them susceptible to fungi – after all, fungi like to breathe, too. A tree that grows quickly rots quickly, and therefore never has a chance to grow old. Because of the extra nutrients in the air this whole process is now accelerating rapidly. The trees are like high-performance athletes already doped up on steroids, who then have an extra dose jabbed into them for good measure.

Luckily, the high nitrogen load in our environment doesn't need to be a long-term problem – provided we can find a way to end our emissions. There are armies of bacteria in the ground that get their energy from once-prized and now harmfully over-abundant nitrogen compounds by breaking them down into their original components. When they do this, the gaseous form of nitrogen escapes from the earth and returns to its original home – the atmosphere – while another portion is washed down deeper into the ground by rainfall, where it pollutes our supplies of drinking water, spoiling our thirst for our most important form of sustenance. As soon as our interference in our ecosystem is reduced appropriately, there's no question the pendulum can swing back. And then, one day, salmon and bears will be in charge once more.

The full effect of their double-act only plays out along streams and rivers, but another force of nature makes its presence widely felt everywhere, shaping mountains, forming valleys and wetlands and, most important of all, functioning as a gigantic redistribution mechanism. That force is water.

3

Creatures in Your Coffee

WATER NOT ONLY CONVEYS NUTRIENTS into the forest through the vehicle of migrating fish, but even more importantly, thanks to its innate properties and the law of gravity, it also carries huge quantities out again. We all know that water flows downhill. But there's more to this seemingly mundane process than meets the eye. I'm talking about the survival of whole ecosystems.

Let's start by looking back into the past. All life on this planet needs nutrients – minerals and compounds that contain phosphorus and nitrogen, for example. Nutrients dictate the vigour with which plants grow, and all animals depend on plants for food. I'm not talking about salmon this time, but us. Our ancestors experienced how inextricably bound up in these cycles of life they were when they cut down forests to make room and building materials for settlements, and then farmed the ground they had cleared.

At first it all worked fine, thanks to the many tens of thousands of tonnes of carbon dioxide stored per square kilometre of ground in the form of humus. But then this soft brown substance slowly began to decompose. Without the cooling shade of trees, the ground warmed up, and bacteria and fungi became active in the soil even deep below the surface. In the orgy of consumption that ensued, not only was carbon dioxide exhaled into the atmosphere, but nutrients that had previously been bound up in the soil were also released. At first the over-fertilisation of the soil that resulted was welcomed: abundant yields ensured people were well fed even when other resources were scarce, and they continued for a few golden years until the fertility of the soil gradually diminished. As there were no synthetic fertilisers at the time and the paltry number of livestock produced only meagre quantities of manure, eventually nutrient levels in the fields were depleted.

The soil still had enough nutrients in it, however, for grass to grow. And so fields previously used to grow crops were used to graze livestock instead. But even now nutrients were still being removed from the land, because the livestock didn't remain in the fields: they were led away to be slaughtered and eaten. The land therefore continued to lose its vitality. Heathers and junipers – plants that sheep and goats don't eat – claimed more and more space. Eventually, all that was left were ruined fields that contributed next to nothing in the way of food. These days we find this kind of landscape romantic: on a summer's day there's nothing quite like a belt of juniper or an expanse of heather dotted with sheep. For our ancestors, however, the sight of juniper berries or heather in bloom presaged destitution.

After the invention of synthetic fertilisers, large expanses of heathland were brought back into agricultural production, because now farmers could spread as many nutrients as they wanted. The few small areas that remained as a testament to old-time agricultural mismanagement are preserved and maintained to this day, but that's another story. What our ancestors did was participate in a grand experiment in speeding up time. They accelerated the natural emission of nutrients, and inadvertently demonstrated what happens when there's no system for replenishing them.

Not that I long to return to the days before fertilisers, because that would mean going back to those same primitive cycles all over again, and my father spelled out to me what that used to mean. After the war, his family kept a vegetable patch as an important extra source of food. As manure was scarce, they spread the patch with the contents of the septic tank. By the time this domestic fertiliser returned to the table in the form of salad greens and cucumbers, nature had enriched it with an additional blessing: intestinal worms. Along with the nutrients, these worms had been recycled from the toilet to the garden and back onto the table. But not even such unsavoury consequences can prevent this kind of nutrient cycle from slowly drying up. Which brings us back to water.

Water is a solvent, and it dissolves all the important substances plants like to suck up with their roots. Yet even though nutrients are taken up out of the ground in this way, when plants die and bacteria and fungi break them down they return to the ground. At least, that's the simple version.

In the normal course of events, moisture seeps deep into the earth until it reaches groundwater, taking with it on its way down all those vital substances that Trees & Co. would so love to keep for themselves. (That, by the way, is why our drinking water needs to be increasingly chlorinated – because the liquid manure that's spread on fields and pastures in unbelievable quantities also ends up, along with its generous portion of bacteria, deep down in aquifers, and therefore in our most important form of sustenance.)

This natural vertical trajectory is of utmost importance for the ecosystem beneath our feet. Numerous creatures that live deep underground depend on scraps from the table of life above the surface. Before we turn to these beings, however, we need to look at the destructive power of water. Not every shower of rain seeps gently into the porous soil of the forest floor to replenish the groundwater. In heavy storms, pores in the soil fill up, and the soil's natural vertical channels overflow. When the ground is saturated by heavy rain, brownish run-off flows into the nearest stream, carrying with it a great deal of organic matter. You can see this for yourself every time you take a walk in the rain. As soon as the rivulets in pastures and fields become murky, they are carrying off soil – valuable soil that won't get replaced for a very long time. Sooner or later, as more and more soil is washed away, the ground gets worn out.

Or at least, that's what would happen, except that luckily nature has stepped in to stop this process of erosion. Nature's main line of defence is the forest. Trees slow the downpour by intercepting a great deal of the rain in the forest canopy. After the shower has passed, the trapped moisture then drips slowly down to the ground. And that's why we Germans have a saying that in the forest it always rains twice. This leafy interception ensures that even heavy rainfall is spread out over a wide area and reaches the ground slowly, giving the soil time to absorb almost all the moisture. Soft moss on tree trunks and ancient stumps does its bit by soaking up the rest. These green cushions can store many times their weight in water, which they gradually release back into the surrounding forest. And because there's hardly any erosion when rain is slowed down like this, the soil layer in ancient forests is usually very porous and deep, and acts like an enormous sponge to absorb and store large quantities of water. In this way, healthy forests create and protect their own reservoirs.

Without trees, the situation changes dramatically. Although grasslands can reduce the impact of heavy rainfall to some extent, ploughed fields have no protection against raindrops pelting down. The fine crumb structure of the soil is destroyed, and the pores fill with mud. Many of our crops, such as corn, potatoes and turnips, cover the ground for only a few months, meaning that for the rest of the year fields are left completely unprotected from the weather, something nature did not anticipate in these latitudes. When a cloudburst hammers the ground, barely any water seeps downward; instead, floodwater streams across the surface.

'Flood' is no exaggeration. A heavy storm cloud can rain down 30,000 cubic metres of water per square kilometre – in just a few minutes. If water is not channelled appropriately – if the rain is not slowed down by foliage so it can seep into open pores in the soil – raging torrents quickly form and carve deep grooves in the muddy fields. The steeper the slope, the faster the torrents, and the more soil they wash away. A gradient of just 2 per cent, which to us looks flat as a pancake, is sufficient for soil to be lost, and the loss is dramatic.

Have you ever wondered why archaeological finds always have to be dug up out of the ground? Shouldn't they be lying around on the surface, covered over by grass or undergrowth? And mountains – created when continental plates collide and thrust upward at the point of impact, a process that continues unaltered to this day – why aren't they getting higher all the time?

Mountains are not constantly increasing in height for the same reason that Roman coins are usually found buried deep in the ground: erosion. Land is higher than the ocean (another blindingly obvious fact), and rain clouds formed over the ocean provide the land with a constant supply of water. That water flows downhill and arrives, sooner or later, back where it started, picking up particles as it goes, and imperceptibly scraping soil off the mountains. The steeper the terrain, the faster the water flows, and the more extreme the abrasion. Our landscapes, however, are not formed by steady, normal rainfall and peacefully babbling brooks, but by rare extreme weather events. When rain pours down for weeks, turning small streams into raging rivers, then the mountains really take a battering. The resulting floods can shift even large rocks, and carry away so much soil that the murky water turns a light shade of brown.

Once things have calmed down again, you can see new contours of the riverbanks where the water has undercut the river's sloping sides with particular force. As the river elsewhere returns to its regular course, the receding floodwaters deposit a thin layer of mud over the rest of the valley. The mud is a combination of water and dust, and the dust is abraded rock: bits of the mountain have been washed down into the valley. Valleys are fertilised by this sediment-rich floodwater – the Nile is a prime example. The highly developed civilisation of ancient Egypt was only made possible because fertile riverbanks allowed farmers to produce a lot of surplus food, and plenty of food on the table means plenty of time to invest in other activities.

But let's return to the forest. What's happening here is that Peter is being robbed to pay Paul, and in this case, both Peter and Paul are trees. Trees often grow high up in the mountains, despite their fondness for nutrient-rich soil many feet deep. But the higher the elevation, the steeper the slopes, and therefore the more severely the soil erodes. And that's why trees on upper slopes don't grow as tall as trees lower down. These trees struggle mightily to fortify themselves against the forces of nature, and over time every crumb of soil they manage to hold on to makes a difference. Just a millimetre of erosion amounts to the loss of almost 1,000 tonnes of soil per square kilometre. Agricultural fields in central Europe lose an average of 200 tonnes of soil per square kilometre per year; that's a loss of 2 centimetres every hundred years.

In extreme cases, however, up to 50 centimetres of soil can disappear over that period, and in the forest I manage I can see the long-term consequences. There's a small hill here with an ancient beech forest on one of its slopes. Despite the steepness of the slope, on this side of the hill there is 2 metres of rich soil. I know the exact depth because this was where, to protect a stand of ancient trees from logging, we established a cemetery – our Final Forest. And to do that, we had to determine the soil's 'inter-ability', as the regulations required. Or, to put it in laypersons' terms, was it possible to bury urns here 80 centimetres down? We hired a geologist to find out and, much to our astonishment, he came across this thick layer of soil. 'This forest must have been here a very long time,' he observed – probably since the arrival of the beeches about 4,000 years ago.

Over on the other side of the hill, however, the bare rock is exposed in places. Virtually all that once-deep soil has gone, leaving just a thin layer of no more than a couple of inches. Back in the Middle Ages, pastoral farming must have been carried out here, and even though grassland suffers less from erosion than fields, the consequences were still disastrous. A few millimetres of erosion over the ensuing centuries added up to metres of soil lost and washed down into the neighbouring Armuthsbach.

We now had a better idea of how the stream got its name. Without soil or humus, the fertility of the land was drastically reduced, and the result was famine. As recently as 1870, people here were dying of malnutrition, and food had to be brought on covered wagon trains from Cologne to the suffering villagers. These wagon trains were regularly ambushed by outlaws: it was like the Wild West out there. And all this happened because the forests were felled, which then led to the almost imperceptible but nevertheless inexorable erosion of the soil.

Can the process be reversed? Yes, it can. This is reassuring news, even if it may take as long as it did to cause the initial damage. Let's assume that the damaged ground is reforested one day and erosion effectively stops: then the layers of soil will begin to build up again. As soon as the rate at which the soil is being eroded drops below the rate at which new soil is being created, the amount of brown gold will increase. The source of this new soil is rock, which is constantly being weathered slowly into tiny pieces. In the conditions we have in central Europe, on average between 300 and 1,000 tonnes of rock are transformed into soil per square kilometre every year. That means an increase in soil depth of 0.3 to 1 millimetres, which would work out on average at least 5 centimetres per century. It would take about 10,000 years for the rocky slope on the hill by the

Armuthsbach to return to the state it was in before the trees were felled and the land was turned over to agriculture, and that is the length of time from the last Ice Age to the present day.

Does that sound disturbingly slow? Well, nature takes its time, as you can see if you consider the growth rate of trees. The oldest spruce in the world, in the Swedish county of Dalarna, is almost 10,000 years old. Looked at this way, one generation in tree terms is all it will take for everything to get back on track.

In our examination of ecosystems and their interconnections, we've already had a good look at what's happening on land. But wait a minute: that's not quite correct. Above ground, yes – but what's going on below ground? The world is a three-dimensional place, after all, and the strata beneath our feet play host to significant ecosystems. I don't mean the 2 metres of arable soil we've just been talking about: I'm talking a whole lot further down. For bacteria, viruses and fungi have been found up to 3.5 kilometres below ground, and even if you go down no further than 500 metres you can find many millions of life-forms like this per cubic centimetre of matter. In these lightless depths, oxygen no longer plays a role in breathing, and in many cases food consists of the materials we're accustomed to using in industry and transportation: oil, gas and coal.

Life in these hidden ecosystems has barely been explored, and we only know about a tiny fraction of the species that live down here. According to the first rough estimates, the subterranean rock layers could be home to 10 per cent of the earth's total living biomass, and because they are so far down and mostly inaccessible, we can assume that, apart from a few coal mines and deep strip mines, these layers have been spared the huge disruptions caused by human activity.

These depths hide another distribution system, however, with which humans have already begun to interfere: groundwater. Groundwater is a very special habitat. Not one single ray of light ever reaches down here, and neither does cold. Depending on the depth, it's either pleasantly warm or extremely hot, and there's not much to eat. At times of climate change, ecosystems like this have a distinct advantage: down here nothing changes.

Despite the lack of food, there's some lively activity going on beneath our feet. OK, maybe not that lively, because – at least in the layers close to the surface, where the temperature sometimes drops below 10 °C – it's not particularly warm. Low temperatures and not much food mean animals slow down. When you get down between 30 to 40 metres below the surface, the temperature rises to 11–12 °C, and it increases by 3 °C every 100 metres you descend. But if you think life is lived at a faster pace down here where it's warmer, then think again.

Topping the charts of the slowest creatures ever is one of the most happy-to-reproduce life-forms in the world: bacteria. Whereas many members of this group reproduce at a breathtaking rate (in our gut, for example – some species of bacteria divide, which is to say double their numbers, every twenty minutes), those inhabitants of the layers buried a kilometre or more below the earth's surface are, it appears, immune to the pressures of time. Some species, it was reported in *Der Spiegel* online after a meeting of the American Geophysical Union in the autumn of 2013, take 500 years to divide.[1] In such conditions, food doesn't spoil and bacterial diseases don't break out, because the hosts (us) would be dead long before these minute creatures had even begun their work. The slow pace of life is down to the inhospitable conditions: at this depth, high pressure and intense heat are the norm. The current record holders manage to survive 120 °C and still carry on dividing merrily – at their own pace, of course.

It seems as though this far down little changes even over the course of centuries. But that isn't quite the case: everything down here is actually in flux. Following heavy rain, water seeps down from ground level. At least it does in central European latitudes, where more fresh water falls from the sky every year than evaporates into the air again. If less rain fell, central Europe would become a desert, and in some regions, as a quick look at the numbers makes abundantly clear, it wouldn't take much to tip the balance. In Germany, an average of 481 litres of water evaporate per square metre every year.[2] In some areas of the state of Brandenburg these days it hardly ever rains at all, which means, quite simply, that the groundwater there is not being replenished. Continuing climate change will only see that rate of evaporation increase, and it probably won't be long before the subterranean habitat will finally be cut off from resupply from above. And it needs to be resupplied, because a small amount of moisture is forever being lost here and there.

Freshwater springs are groundwater's gaping 'wounds'. What we see as merrily bubbling wonders of nature are for some of the denizens of the deep an absolute catastrophe. Water forced to the surface through layers of rock rudely washes crustaceans and worms up into the light of day where, thanks to the abrupt change in their environment, they quickly expire. Winter is a particularly good time to notice these upwellings of groundwater, because they don't freeze where they bubble up out of the ground. (Groundwater maintains a constant temperature of about 10 °C, which drops only when it encounters fresh air and everything around is frozen solid.) So if you find water that's still gently moving even when temperatures outside are well below freezing, you can be certain that water is coming from deep underground.

But let's get back to biodiversity. According to recent research, groundwater harbours a surprising wealth of crustaceans and other minuscule creatures. They paddle blindly through dark currents and, every once in a while, they probably end up in the water you use to make your morning coffee. Most treatment plants pump water into their reservoirs from deep underground, tapping into what until their intrusion was essentially a hermetically sealed habitat.

Tiny creatures in your coffee despite the elaborate filtration procedures at those water-treatment plants? Yes, indeed: despite all efforts to keep them out, pesky little creatures like water lice (which can grow to almost 2 centimetres long) often make it through to live happily in the water pipes on the other side of all those purification systems. After all, the water pipe in your basement is basically an extension of the groundwater system – it's dark, cool and clean down there.

You'll be aware of this the moment you turn on the cold-water tap: the water that comes out is the temperature of groundwater. As soon as it does, some of these little scoundrels can't cling on any longer and are swept along in the flow – and end up, via your cup of coffee, in your stomach. But water lice aren't the only creatures in the water system: many others are much smaller. Bacteria, for example, form a thick layer that coats the inside of these metal pipes. And in every sip we take there will be traces of them too.

You can look as hard as you like, but you won't be able to see most of these uninvited guests (with the exception of such giants as water lice), at least not without the aid of a microscope. In the absence of light, there's no point having eyes or colour, which is why as a rule these groundwater dwellers are blind and transparent. The lack of light, however, brings up another problem. No sunlight means no photosynthesis, which means no food being produced in the form of plants. These subterranean multitudes are reliant, therefore, on handouts from the biomass of plants and animals up above that decomposes into humus and, along with the rainwater that seeps into the ground, sinks slowly into the depths.

On the way down, these nutrients are metabolised many times over, because there's an entire food chain down here, just as there is above ground. Most of it is made up of bacteria, which colonise everywhere and form layers, just as they do in water pipes. These bacteria are in turn browsed by minute predators like flagellates and ciliates. It's a good thing these voracious Lilliputians exist, because if they didn't, the pores of plutonic rock deep below the surface would eventually become clogged. But these tiny predators meet their match as well in the form of heliozoans, also known as sun animalcules. They're slightly bigger, and particularly enjoy eating their fellow creatures.[3] And so there is a complete ecosystem underground of which we are only dimly aware, except where we pump its (and our) elixir of life – water – up to the surface to satisfy our needs.

Speaking of which: we got sidetracked while we were talking about our morning coffee and contemplating the stowaways in our mug. If you're disgusted by the thought of bacteria in your beverages, perhaps this extra titbit of information will make you feel better: you yourself are the mother ship for these tiny creatures. Apart from the 30 billion cells that make up your body, you are also host to around the same number of bacteria, most of them in your gut.[4] Thousands of different kinds of bacteria are floating around inside you, and in most cases, they're vital for your survival – helping you fight off illness, for example, or break down foods that are hard to digest. Does it really matter if a few more harmless fellows find their way inside you via your drinking water, especially as they won't survive for long in your digestive tract?

Forests are important for groundwater, to such an extent that in Germany some water companies even make bonus payments to forest owners in return for pursuing responsible management policies. This doesn't seem to make much sense, though. For one thing, trees are heavy consumers of water. On a hot summer day, a thirsty mature beech can suck up to 500 litres of water out of the ground. It uses this water in a number of different ways, but most of it ends up evaporating out of its stomata (the tiny openings on the underside of its leaves). Grass would use a lot less.

But trees, especially deciduous trees native to central Europe, have an advantage over grass, because they also collect water. They gather rainwater with their upward-spreading branches, which funnel the water to the trunk and down to the roots. During a heavy storm I once stood under an ancient beech – don't try this at home! – and observed this water collection for myself. So much of it cascaded down the trunk that it foamed up around the base of the tree like beer fresh from the barrel.

Once the water reaches the ground, it soaks into the loose soil, which is as absorbent as a sponge. Even heavy downpours soak in and slowly trickle down through the soil layers. When there's no more rain trees will, it's true, in due course help themselves to some of this water – to them the ground around their roots is like a reservoir they can tap into any time they're thirsty – but the rest seeps down to layers too deep for the roots of plants to reach. And at this depth, this water slowly becomes part of the flow of groundwater.

Where I live, resupply of groundwater only takes place in winter, when the plant world is hibernating. Beeches and oaks take a break for a while, and the water can slip unchecked past the tree roots and down to the depths. In summer, on the other hand, there's never enough rain to satisfy the trees' thirst. They greedily suck all the moisture out of the ground and pump it into their trunks.

In these times of climate change the manner in which trees take up water gives me pause for thought. Warmer temperatures will change many of the parameters. Water will evaporate more quickly, which means the ground will dry out sooner, even without plant activity. In addition, trees, like us, drink more in hot weather. And longer growing seasons will shorten the times when trees take a break and forests hibernate, and the ground can recharge its water supplies. But even with these caveats, forests should still be able to generate enough new groundwater beneath them in future – as long as we don't damage them too much with our logging operations.

Open grassland and especially agricultural fields, however, are less good at absorbing rainfall. The ground is compacted by wild or domesticated grazing animals, although these days the major cause of soil compaction is heavy agricultural machinery, which compacts the soil much further down than hoof or trotter. Compressed in this way the sponge-like earth never, unlike the foam pan-scourer in your kitchen, regains its original absorbent structure. That in turn means heavy rain can no longer be absorbed. Instead the water runs off downhill in rivulets with increasing speed to end up in the nearest stream (which leads to the nearest river, which carries the fresh water out to the ocean). And so that water is lost to the local groundwater supply, and the whole process speeds up erosion.

Add to this the fact that the air heats up much more quickly above pastures and fields than it does over forests, which means the ground dries out more quickly and life-giving moisture escapes into the air, to be carried away, only intensifying the effects of desiccation.

The greatest danger for groundwater, however, is not climate change but the extraction of raw materials from the earth, especially through fracking. Here water is pumped deep into the ground under high pressure to fracture rocks. Grains of sand and chemicals mixed in with the water hold the fractures open, which allows the gas and oil contained in these rocks to flow to the surface. But the underground ecosystem is not equipped to deal with such brutish intrusions. In this habitat, as we've seen, hardly anything ever changes, and those changes that do take place do so extremely slowly. One can only hope that not too many areas are opened up for this kind of mining.

If we manage to avoid fracking, the best means of protecting groundwater is the forest. The trees in it are the secret protectors of tiny crustaceans living hundreds of metres beneath their roots. Other animals, however, have a more strained relationship with beeches and oaks. You could say that the relationship between trees and deer literally leaves a bad taste in the animals' mouths. And it turns out that deer don't taste good to trees, either.

4

Why Deer Taste Bad to Trees

DEER HAVE A LOVE-HATE relationship with trees. They
don't actually like forests, but we think of them as forest
animals, because most of the time that's where we find
them. Like all large animals that eat plants, deer have a
problem: they can only eat vegetation they can reach.
And usually the vegetation available to them has armed
itself against herbivorous attack. The usual armoury
includes thorns and barbs, toxins or simply thick, hard
bark, but trees in central European forests have developed none of these defences. Does this mean the trees'
offspring have to endure every bite from browsers
without being able to fight back?

If you take a good look around the forest, you'll see
how beeches defend themselves. The forest floor around
deciduous trees is conspicuously empty of vegetation.
Here and there you might find a lonely fern, or a few
grasses in a tiny clearing where an ancient giant has
fallen, allowing a few rays of sunlight to reach the
ground. In general, however, light levels are too low for
plants to produce copious quantities of sugar, which
means that wild plants growing in the forest contain

few nutrients in comparison with their relatives growing out in the open. In other words, understory plants are tough and bitter.

Most of the forest is dark, because only 3 per cent of the sun's light penetrates the canopy. For the plants under the trees this makes it pitch-black. You might not think so when you walk through the forest, but this has to do with the green shade you find there. Trees use the chlorophyll in their leaves to convert light, water and carbon dioxide into sugar. Chlorophyll, however, has a 'green gap', which means it can't make use of this wavelength of light. As a result, green light is reflected, and this makes the forest seem brighter to human visitors than it does to plants, because plants cannot 'see' this colour. As 97 per cent of all the other wavelengths of light have already been absorbed and processed in the canopy, from where the green plants on the forest floor are standing, things literally look gloomy.

This is the case for young beech trees too. The few rays of sunlight falling on their small, thin leaves enable them to produce such meagre amounts of sugar that their twigs and buds contain barely any nutrients. To ensure that the next generation of trees doesn't starve because its opportunity to photosynthesise is so limited, mother trees supply the saplings with nutrient solutions through connected root systems – you could say they suckle their offspring. Non-woody plants and grasses in the forest don't benefit from this kind of assistance, however, and can therefore only grow in the tiny clearings opened up when an old tree falls.

This is what the supposedly idyllic forest looks like to roe deer, then: a few patches of dry, stringy grasses and non-woody plants scattered among tough young beeches. Even if the beech leaves were halfway palatable, they would make a very boring diet, which deer don't enjoy any more than we do. Imagine having to eat your favourite food day in, day out for months on end – you'd soon lose your taste for it. Deer prefer to avoid monotonous meals that don't offer much variety in either taste or nutrients, especially when they have to produce milk for their fawns. They find forest-edge habitats much more attractive. Here, maybe along a riverbank, grasses and non-woody plants grow in full sun in fertile soil until they positively brim with energy. Unfortunately, in forest-rich central Europe, few such edge habitats occur naturally, which is why historically forests here have had only very small populations of roe deer.

It's no surprise that roe deer prefer disturbed areas. When a summer tornado uproots a small stand of ancient beeches, an island of light opens up in the forest. The plants I've just been describing, that struggle to survive in the shade of the forest, soon colonise the clearing. And they have a lot to offer deer. Bright sunlight means full-on photosynthesis, which means tasty carbohydrates in their leaves and buds. Even the beech saplings, which now find themselves unexpectedly illuminated, become sweet and tasty. And now, in this clearing, is the paradise this small species of deer has been searching for. Roe deer love food with a high-energy content: scientists call them concentrated feeders. If we were to eat in the same way, our meals would consist of nothing but fast food and chocolate enriched with vitamins. Roe deer don't have to worry about getting fat, however, because such calorie-stuffed oases rarely occur in nature.

If you're a small herbivore, it's a bad idea to run in the face of danger. Wolves could easily follow you and attack, so it's better for you to hide. Roe deer don't run very far before they double back and try to return to their original location, and when they do, they cross their own tracks, which confuses their pursuers – which trail should they follow? Once they're safely back on home turf, roe deer hide in stands of small trees. And because herds are easier to spot than single animals, roe deer live alone. But another reason for their solitary existence is the lack of food in ancient undisturbed forests. A herd of deer would have to cover a lot of territory to find sufficient food. Travelling long distances, however, increases the risk of coming across a pack of wolves. And so the single life is better.

The need to be solitary also means that a mother leaves her offspring behind when she goes out in search of food. This is perfectly normal behaviour in the first three to four weeks after a fawn is born – the time when the little one can't yet keep up. To make sure her fawns (usually she has twins) don't slow her down, the doe leaves them lying in tall grass or under a bush. When a predator approaches, they flatten themselves to the ground so they can't be seen. Unfortunately, some people interpret this behaviour to mean that a fawn has been abandoned, and take the supposedly helpless young deer home with them. Here, because it refuses to take milk from a bottle, it often dies a painful death from starvation.

Life without a large extended family is typical for many forest dwellers, including the lynx. Lynx strike out alone across their enormous territory, which can sometimes exceed 100 square kilometres, and seek brief contact with a lynx of the opposite sex only at mating time. By contrast, red deer, which originally lived on grassy plains, behave completely differently. They're social animals that live in large herds, and only go off on their own when the does are ready to give birth – which they like to do in peace and isolation. When predators appear, red deer flee together over great distances in search of a place with good visibility in all directions. They've retained this behaviour despite the fact that human activity – in central Europe, at least – has now forced them to retreat into forests. We don't want to share our open spaces with them, because these are the places where we like to settle and farm.

Back to the roe deer. Life is better than ever for them in Germany now no dark, ancient forest is left. What we think of as forests today are drastically different from forests of days gone by. A bird's-eye view of the landscape – perhaps one of those satellite images you can find on the internet – will reveal what looks like an enormous patchwork with some pieces missing, and these patches – the forest – are small from an ecological perspective: anything less than 200 square kilometres in size is not large enough to support even one wolf pack.

But these many tiny scraps of forest are a real boon for roe deer, because they can find their preferred edge habitats all over the place. Whereas once the collapse of a couple of trees gave them a lucky break, these days the light which non-woody plants and grasses require to proliferate reaches not just the edges but most of the forest floor. Forestry, after all, is simply the practice of growing and cutting down trees. Clear-felling is the most brutal form of harvesting timber, but for browsers it's a windfall. Once the pesky shade from treetops has been removed, non-woody plants and grasses can take over. Not only do the plants now have space and light, they also get a massive boost of fertiliser. The bright sunlight heats the ground so much that even fungi and bacteria deep below the surface warm up enough to spring into action and break down all the humus in just a few years. In the process so many nutrients are released that the growing plant populations can't take them all up. The plants grow quickly and are full of sugar and other carbohydrates, making them tasty treats for roe deer. In areas such as these, the deer don't need to move around much: they can find enough food in a few square metres to keep them comfortably full for the whole day.

Under such conditions their populations explode, because, like all species, roe deer immediately convert food into offspring. Instead of one fawn, there are two, even three, and the ratio of the sexes shifts in favour of females. That heats up population growth even more, which is ideal from the animals' perspective, as it means that roe deer can completely take over the habitat and exploit it down to the last blade of grass.

In Germany there have been noticeable increases in the growth of wild animal populations, especially after the severe storms of 1990 and 2007, which mowed down entire forests. Spruce trees bore the brunt of the damage, but pines and Douglas firs growing in plantations also suffered. The conifers began to tumble when winds blew more than 100 kilometres an hour. The trees fell because their root systems had been damaged while they were still in tree nurseries, being cut back to make them easier to transplant: if the roots are shorter, then you don't have to dig such a big hole for the trees. The flip side is that transplants with trimmed roots never develop an intact root system, which means when storms hit it's almost impossible for them to hold fast to the ground. What makes things worse is that conifers hold on to their needles, making lots of surface area to offer winter storms. This is in marked contrast to beeches and oaks, which pare down their profile by dropping their leaves in the autumn and therefore survive most winter storms unscathed. And this is how it comes about that conifer plantations indirectly favour roe deer.

In the past, in addition to storm damage, the deer also benefited from clear-felling of trees as part of commercial forestry. With clear-felling you harvest a complete stand of trees of the same age in a single pass, which is much cheaper than felling individual trees in thinning cuts. Recently, however, clear-felling over more than a hectare has fallen out of fashion. Too bad for the roe deer? Not at all, because thinning cuts are almost as beneficial for vegetation on the forest floor. Individual trees are removed regularly to give particularly good specimens room to grow. Constantly removing trees in a regular rotation in effect amounts to a more widely distributed, less intense way of clear-felling. In contrast to an untouched ancient forest, trees today make up less than 50 per cent of the biomass of a cultivated forest. This means that more light reaches the ground, non-woody plants, grasses and bushes take hold over large areas, and the lower storeys of the forest have warmed up (by about 3 °C). The resulting deer buffet is not as lavish as the feast laid out in the clear-felled areas; however, it makes up for this by being available almost everywhere in the forest.

As about 98 per cent of the forested areas in Germany are cultivated, that's like having a gigantic nationwide feeding programme for deer. Add in the hunters who put a great deal of effort into caring for their potential prey by hauling tonnes of feed into the forests, and you can see why deer populations are increasing by leaps and bounds. Today, as I mentioned earlier, there are fifty times more roe deer wandering around our forests than there were before the advent of these intrusive practices.

You can easily see for yourself where and how forest landscapes have changed. With the exception of a few tiny areas, grasses, non-woody plants and bushes hardly ever grow in natural forests in our latitudes. Large-scale coverage by these plants can always be traced back to human interference in this ecosystem, which for roe deer at least is cause for celebration.

Some of the plants see things differently, however, because when it comes to food, this little species of deer has preferences, just like us. At the top of the list are saplings of beech, oak, cherry and other deciduous trees, along with the offspring of the now-rare white pine. After that, the metre-tall stalks of fireweed topped with spikes of radiant magenta flowers and the somewhat inconspicuous wild raspberries are perennial favourites. These are the first delicacies to be eaten, and if roe deer populations are high, these plants disappear completely, allowing others better able to defend themselves, such as blackberries, thistles and stinging nettles, to take over.

It's obvious that trees native to the ancient forests of central Europe never had to deal with browsers: they have developed next to no real defences against hungry mammals – no thorns, no toxins in their leaves, no impenetrable tangle of branches. It seems beeches and oaks offer their almost defenceless seedlings to any animal that might want to take a bite. Their only protection is the eternal twilight of the forest floor and the aforementioned lack of plants, which make the forest an unattractive place to live.

But these flimsy defences only work when there are very few animals around, such as roe deer. They'd be no use against large herds of hungry aurochs or tarpans (ancient horses), which would simply have torn the bark off the trees. The dying trunks and crowns would have created space and light for grassy plains to become established, and the herbivores would then have been able to nourish themselves from plants that grew out in the open – and the forest would have disappeared. But none of that happened in central Europe. That, it seems to me, is a clear indication that there never was a serious, lasting threat from these animals, or evolution would have come up with counter-measures.

Things are quite different for plants adapted to life on grassy plains. Wild horses, wild cattle and red deer are all at home in these wide-open spaces, and love to nibble on the fresh growth of trees and bushes to vary their diet. Woody species that grow in such surroundings defend themselves vigorously against their attackers. Blackthorn is a classic example. The dagger-like thorns – even on bushes that have been dead for years – are sharp enough to easily pierce not just skin (of any kind) but also rubber boots and car tyres. The wild apple deploys similar defensive weaponry. Like the blackthorn, it also belongs to the rose family. Indeed, their thorns identify members of the rose family as native grassy-plain dwellers.

Plants that prefer not to equip themselves with thorns depend on toxins. Among these are foxglove, broom and ragwort. This last is especially dangerous because its harmful effects accumulate over time. First it causes mild liver damage, but at some point the animal eats one plant too many and it dies.

Not every species, however, is affected by the ragwort's poison. There are butterflies and moths that not only consume these pretty yellow-flowering plants but also use them for their own protection. The cinnabar moth is one such. Its caterpillars happily chew through one tiny leaf after another all day long, taking in not only the calories the plant provides but also its poisons. The caterpillars suffer no ill effects whatsoever, but the predators that eat them do, and so the caterpillars sport black and yellow rings to warn attackers that they will make a deadly meal. This colour combination seems to be a universal warning in the animal kingdom: just think of the yellow and black on wasps and salamanders, for example.

Everywhere you look, plants struggle not to be eaten. Indeed, recent research has shown that, although they go about it very quietly, deciduous trees are not as passive as we (and I) long supposed either. To investigate this further, scientists at Leipzig University and the German Centre for Integrative Biodiversity Research (iDiv) simulated attacks on small beeches and maples. Whenever a roe deer takes a hearty bite out of the top growth of a young tree, it leaves a little saliva behind in the wound, and it soon became clear that wounded trees can clearly detect the presence of this saliva. To simulate browsing by roe deer, researchers cut off buds or leaves and dripped roe deer saliva from a pipette onto the damaged areas. What they noticed was that in response the little trees produced salicylic acid, which in turn led to an increased production of bad-tasting defensive compounds, which discouraged the roe deer from eating them. However, when the scientists simply broke off new growth without applying any saliva, all the beeches and maples produced were hormones to heal the damage as quickly as possible.[1] This experiment also proved that these trees (and possibly many other species) can 'taste' the saliva left on leaves and shoots and recognise that they are being browsed by herbivores.

After the deer population reaches a certain density, however, knowing what's eating you doesn't help. The animals polish off so many of the plants in their territory that they will even nibble all the bad-tasting leaves off beech saplings. Forest owners at their wits' end try to help the little deciduous trees by smearing bitter-tasting concoctions on their leaf buds. In my early years as a forester I tried this myself, but the roe deer soon put paid to it: they were so hungry they just ate the white paste along with the buds.

Forest floors scoured of vegetation – and the ageing of the forest that comes with it – are an acute problem in many parts of central Europe, and show how wild animals have reached population levels the trees have never had to deal with before. How might we change this dynamic in future? One way would be to leave more trees in the forest. In other words, foresters should step back. Allowing more trees to grow would make the forest darker once again, and that, in turn, would allow beeches and oaks to employ their tried-and-tested strategy of light starvation. And if hunters also gave up their winter feeding programmes, the situation would improve considerably. If the wolf were to arrive as well (and it's on its way), perhaps we'd eventually have a situation similar to Yellowstone right here where I live.

None of this would make the clockwork mechanism of nature completely regain its former rhythm, because no one can or will remove the patchwork of pastures, agricultural fields and smaller forested areas that covers the landscape of central Europe – not even me. After all, I'm hungry every morning too, and enjoy biting into my breakfast roll as much as anyone else, and for that I need someone to farm a wheat field.

But it's not only roe deer that profit from our trans-formation of the landscape to suit our own purposes. There are other brown-coloured creatures that have a great deal of influence on the environment in which we live: they are teeny-tiny, really good at defending them-selves, and they have a weakness for forget-me-nots.

5

Ants – Secret Sovereigns

ALL SUMMER LONG, our yard is full of forget-me-nots. These clumps of blue pop up here, there and everywhere, and are always wandering uninvited into our vegetable beds, where they make themselves at home and refuse to leave. And because they are so pretty, we usually leave them alone and accept their intrusion. Forget-me-nots, however, can only conquer new territory so successfully because they have an army of tiny allies: ants.

It's not that ants are particularly fond of flowers – at least, they're not attracted by their aesthetic qualities. Ants are motived by their desire to eat them, and their interest is triggered when forget-me-nots form their seeds. The seeds are designed to make an ant's mouth water, for attached to the outside is a fleshy structure called an eliaosome, which looks like a tiny cake crumb. To an ant this fat- and sugar-rich morsel is like crisps and chocolate. The tiny creatures quickly carry the seeds back to their nest, where the colony is waiting eagerly in the tunnels for the calorie hit. The tasty treat is nibbled off, and the seed itself is discarded. Along come the trash collectors, in the form of worker ants, which dispose of the seeds elsewhere in the neighbourhood – carting them up to 70 metres away from home. And it's not just forget-me-nots that get a helping hand. Wild strawberries and wood violets also benefit from this distribution service: ants are nature's gardeners, as it were.

There's a huge army of them pottering about in forests and fields, and in some respects they are as busy as we are. So far about 10,000 species of ant have been discovered, and the German newspaper *Die Zeit* once tried to estimate the combined weight of all the tiny creatures in this family of insects. According to its calculations, it's equivalent to the weight of all the people on earth.[1]

Although most wood ants are small, their colonies and the structures they build are often very large. The largest anthill I've ever found in my forest was almost 5 metres wide. My first experiences with red wood ants (the most common forest ants) were on childhood walks with my family. As soon as one of us spotted a sizeable hill of these social insects by the side of the path, the ritual was always the same. My mother tapped the outside of it gently with the palm of her hand. When she held her hand under our noses we could immediately smell a pungent, sour odour. The ants had turned their backsides towards us and sprayed acid to fend off the intruder. Throughout this demonstration we had to hop quickly from one foot to the other to make sure none of these feisty little creatures ventured over our shoes and up our trouser legs to bite us – their bites are extremely painful.

It's hardly surprising that wood ants are so good at defending themselves: after all, they are related to honeybees, and share a similar social organisation. One difference is that ant colonies can have multiple queens; another is that related ant colonies tolerate each other, which can't be said of honeybees. The latter raid each other's hives, especially in the autumn, and the defeated colony is mercilessly dispatched and its hive emptied of its honey. Ants are more peace-loving – at least when it comes to dealing with their own kind: they like other insects only in a culinary way. Adult bark beetles and larvae, for example, are eagerly carried away and fed to the ant larvae back at the colony. They're so insatiable that during the summer, millions of beetles in a radius of up to 50 metres or so from the hill end up as ant meals.

In spruce plantations, the dreaded spruce engraver beetles feed. In large monocultures of pines, it's the larvae of pine-tree lappets and pine beauties that strip whole forests bare. Not, however, in the vicinity of red wood anthills, within range of which islands of green remain amid an ocean of dead trunks. This led to red wood ants being dubbed the public health patrol of the forest. Henceforth, the ants working so hard on behalf of foresters and forest owners came under strict protection, for red wood ants eat not only the aforementioned pest species but also carrion, which makes their honorary title even more appropriate. Though not by design, the

new rules also benefit rare species of bird that dine on
ants. The crow-sized black woodpecker, along with black
grouse and wood grouse, or capercaillie, like to help
themselves to larvae and pupae from anthills. Red wood
ants, therefore, can clearly be categorised as beneficial.

But if we take another, closer look at the species,
mild doubts arise – indeed, the question of whether these
ants are actually worth protecting in the first place. Let
me be quite clear: every species, regardless of whether
it is common or rare, is worth protecting, in the sense
of showing it respect. However, protection in the sense
of mandating active support is something quite different,
and in the ants' case it's misguided, at least where I live.

Red wood ants arrived as a result of human interference, and they are able to expand their range only
because of the unrestrained growth of conifer plantations. They weren't around in the original deciduous
forests of central Europe. Have you ever seen a red wood
anthill made out of leaves? No, these ants only use
needles. Moreover, in the spring they need a lot of sun
to be able to start work. The ants move to the surface
of their hill to heat up in the sun, and then crawl back
inside to radiate the warmth they've gathered. Sun that
reaches the ground would have been a scarce commodity
in our original beech forests – yet another strike against
these tiny civil engineers.

But even in their natural habitat, it's questionable
whether the impact red wood ants have on the trees is
entirely beneficial. They're no doubt happy that the
insects remove the bark beetles that attack them, but

the ants' diet is not restricted to meat. It includes sugary foods as well. And in the forest, sugary treats come almost exclusively from aphids. Aphids attach themselves to the trees' needles and bark, stick their mouths down to where the trees' sap flows, and tap into the trees' life-blood. Thanks to photosynthesis, this 'tree blood' has a high sugar content, but that's not what the aphids are after. What they want is protein, which is found in this fluid only in very small quantities. Consequently, the aphids need to allow enormous amounts of the trees' fluids to flow through their bodies to be able to filter out enough of the scarce substances they desire.

Whoever drinks a lot must also excrete a lot, and aphids excrete almost constantly. If you park your car under aphid-infested trees in summer, your windscreen will tell you all you need to know – in just a few hours, it will be covered with sticky droplets. And because the little creatures are constantly eating and excreting, over time their rear ends can get gummed up with sugar. Some species resort to covering their excretions with wax so they can expel them more easily; others enlist the help of ants. Ants lap up the sugary faeces because, like their relatives the honeybees, sugar is the most important component of their diet. Every season, a single ant colony digests about 200 litres of these sugary drop-lets. This so-called honeydew makes up two-thirds of their calorie intake. To give you an idea how much honeydew this is, at the same time an average of 10 million insects weighing a total of 28 kilograms end up in the ants' stomachs to make up most of the remaining third; tree sap and fungal threads only make a small contribution.[2]

Wood ants and aphids, therefore, form a package. And this casts doubt on the term 'public health patrol', because aphids damage trees. First, they sap beeches, oaks and spruce of the energy these trees desperately need for themselves. Then, when they insert their mouth-parts and begin siphoning off fluids, the aphids inflict heavy damage on the trees' tissue. The red-eyed nymphs of the spruce aphid, for example, which are only 2 milli-metres long, drain the needles of many different species of spruce. The needles turn yellow, then brown, and finally drop off. Afterwards, the trees look as though they've been plucked, because only the new season's growth remains on their branches, and their growth is stunted, because the trees' capacity for photosynthesis is now severely limited.

On top of all this, the aphids' activities support pathogens that can be deadly for trees. A kind of aphid known as beech scale feeds on the bark of beeches. These small creatures are covered in soft waxy hairs. They're not dangerous, so long as there aren't too many of them – beeches have no problem healing small individual puncture wounds. The situation is quite different, however, when their populations explode. Aphids have no need of males to multiply, and none has been found. The females lay unfertilised eggs that hatch into larvae. These larvae are carried on the wind to nearby beeches, where they imme-diately begin feeding. When all the nooks and crannies in the bark are occupied with colonies of white scale, the trees look as though they're covered in a light layer of mould, and the defence systems of many become overwhelmed. The tube-like mouthparts of the aphids create weeping wounds that won't heal. The sap that bleeds out is colonised by fungi, which eventually work their way into the trunks and kill the beeches. A good number of trees survive the infection, but they carry the scars on their bark for the rest of their lives.

Loss of life-giving energy and the spread of disease through puncture wounds: for trees, aphids are far from a blessing. And now the so-called public health patrol turns up. The red wood ants could just gobble down the little green pests and bulk up on the protein they provide. However, it's clearly far more advantageous for them to keep the aphids alive and play honeydew farmer. They have to get their 200 litres of honeydew somehow, and what better way than by keeping aphids in the trees around the anthill? In fact, as they defend their herds of aphids against predators the ants win twice: not only do they get to protect their source of honeydew, but they also get access to prey like aphid-loving ladybird larvae, which come to eat aphids but end up getting gobbled up by the aphids' protectors.

Despite the protection the ants offer, however, the aphids are not always content to stick around. When they want to leave, the next generation grows wings so they can fly to greener pastures. This doesn't escape the notice of their guardians, and the ants end the aphids' dreams of flight by summarily biting off their transparent appendages. As if that were not enough, the ants also use chemical means to prevent their domesticated herds from escaping. They exude compounds that slow the growth of the aphids' wings and, for good measure, they also slow down the aphids themselves. A research team from Imperial College London discovered that they move more slowly when they cross terrain that has previously been walked over by ants. The cause is a chemical message left by the ants that affects the behaviour of the aphids and forces them to reduce speed.[3] The beautiful symbiotic relationship between ants and aphids turns out to be not entirely voluntary, after all.

Now you might argue that the aphids still benefit from the ants' attentions: they don't have to worry about being attacked by ladybird or hoverfly larvae, for example. And the 'milking' process doesn't exactly harm them. After all, the sugary droplets that the ants feed on are simply waste products – and carting them away keeps the aphids nice and clean. The sticking point is that the aphids would rather seek out more productive trees when they notice that conditions are no longer optimal in the place where they originally landed. But their protectors, who have now turned out to be their jailers, prevent them from moving away. And these jailer ants that keep their 'livestock' in trees in unnaturally high concentrations are supposed to be the public health patrol?

So are red wood ants really helping foresters when they weaken the trees around their anthills by setting up their honeydew farms? It's not an easy question to answer. At the beginning of this chapter, I mentioned the green islands left in coniferous forests after bark beetles have invaded. No matter how many aphids live in the trees that have been saved, the live trees are still better off than their dead companions. And this brings us to the key to understanding the complicated coexistence of different groups of insects. Trees are attacked not only by aphids and bark beetles but also by a multitude of other species, all of them with one thing on their mind: getting their fair share from the gigantic warehouse of carbohydrates that is a tree. Woodboring beetles lay their eggs on bark and their larvae then tunnel into it. Weevils chew leaves until the edges look as though they've been riddled with shot. This kind of damage is probably much more detrimental to the trees than donating some of their vital fluids to aphids. Red wood ants certainly ensure that there are more aphids around – which means more loss of blood for the trees – but this also leads to more ants in the vicinity. Lots of food in the form of fluids means lots of ant larvae that can be fed. And the more the ants climb up into the trees to hunt the insects that threaten their aphid herds, the fewer attacks from these predatory insects the trees have to endure.

The more interesting question is what the overall ant–aphid–forest balance looks like. Science hasn't yet come up with a definitive answer, although most studies conclude that the positive effects generally outweigh the negative. John Whittaker at Lancaster University, for example, discovered that, on balance, birches did significantly better when there were ants around. Although ants increased the number of aphids, this was true only for some species. Those not farmed by ants declined drastically. Furthermore, leaf-eating insects in general declined so significantly that leaf loss was six times less than in birches not settled by ants.[4] According to Whittaker, plane trees also seem to come out mostly ahead. Aphid-farming ants reduce attacks by other plant-eating insects to such an extent that these trees increase their girth two to three times faster than plane trees that have to manage without the ants' protection.[5]

Does that mean that wood ants are beneficial? The ecosystem is too complex, I believe, for us to be able to answer this question conclusively. If we take this one step further, you will see that trying to understand all the connections here is a Sisyphean task. We could start by asking about sugar. At the end of the day, despite the aphids' bloodletting, a tree can always produce more, because it still has leaves that are no longer being eaten by caterpillars. However, the sugar would normally remain inside the tree, and from there it would end up in the soil ecosystem via the tree's roots and the fungal networks in the ground. But thanks to the multitude of well-tended aphids, sugar from inside the tree now drips down onto the ground and the vegetation below. The ants can't consume all of it quickly enough, so many leftover droplets land on leaves and the surface of the soil. (Remember that car with its sticky windscreen after it was parked under trees.) This sugar is now lost to the fungi that live in a symbiotic relationship with the trees and provide services to their roots: if a lot is squandered above ground, little ends up below ground. Poorly provisioned fungi produce fewer fruiting bodies, which snails and insects rely on for food. Little wonder that it is almost impossible for scientists to evaluate the overall balance.

It's easier to lay out the stark changes caused by commercial forestry practices. By doing away with the original forests – by planting monocultures of trees to harvest for timber – not only are native species displaced (in central Europe, this would be beech), but so are the living communities that depend on them. Earlier we were talking about individual cogs in the mechanism of nature, but here the mechanism itself is being replaced. Whether the new clock will work as well as the old one is open to question.

Unfortunately, the public health patrol does not concern itself with how the clock functions as a whole, but only with a few miscreants. We've already met some of them: pine-tree lappets, pine beauties and bark beetles. Now let's take a closer look at the last.

6

Is the Bad Bark Beetle All Bad?

TYPOGRAPHER, ENGRAVER — BEHIND the delightful names hide insects that top the list of the most-feared trouble-makers in our forests. I'm talking about bark beetles: I'm sure you've heard of them. These days the name has such a negative connotation that I'm frequently asked if all the dead wood in our forest reserve isn't just a breeding ground for pests like this, and whether it would be best to get rid of it. But Engraver Beetles & Co. pose no threat to healthy forests, and happen to be quite wonderful creatures. So let's take a look at them in their natural habitat.

As their name suggests, bark beetles live in forests. They dwell in trees, but not just any old trees. Every species of beetle has its own preferred species of tree. The spruce engraver beetle, for example, which is quite a large fellow, specialises, as you might have guessed, in spruce trees, and is therefore restricted to places where they grow. In spring, when the thermometer climbs towards 20 °C, adult beetles come out from the hiding places under the bark where they've spent the winter and begin their dispersal flight to find a mate. But things aren't quite that simple. The little males have to make elaborate preparations if they want to get lucky.

First, they need to find weakened spruce trees. Like all trees, spruce can defend themselves against insect attack – and who wants to die just before their first opportunity to have sex? Consequently, the beetles narrow their search to trees broadcasting scent signals that betray their weakness, because trees let each other know when they are stressed. For example, if there's a dry spell and a dangerous lack of water in the ground, the first trees to notice can forewarn all their fellow trees in the area. These can then pre-emptively reduce their water use to make the remaining supplies in the root zone last longer. Unfortunately, the trees' enemies also pick up on the fact that someone is in danger of running dry. Normally, spruce defend themselves against insects trying to bore into them by pushing out blobs of pitch to drown them – pitching them out, as it were. But if the trees are short on water or weak-ened in some other way, they don't have enough energy to produce pitch.

Once the male engraver beetle has found a likely candidate, it immediately begins boring into it. 'All or nothing' is the beetle's motto, and if the little male is lucky, no pitch comes out of the tunnel it drills and the gamble has paid off. It continues to make its way beneath the bark, excavating a tunnel that runs parallel to the bark fibres. It advances, millimetre by millimetre, backing out to expel the sawdust it creates.

This brownish frass is a red flag for foresters, because it's a clear indication that the spruce tree can no longer defend itself and is doomed. Once the beetle has succeeded in getting this far, it sends out a scent signal to summon more of its colleagues. It doesn't seem to make sense to broadcast an invitation to other males at mating time, but there's method in its madness. A brief rain shower might be enough to re-energise the tree sufficiently to be able to quickly dispatch the courageous pioneer with a fresh dose of pitch. And so the spruce must be weakened quickly until there's no way it can recover. The more insects that bore into it, the more surely they can extinguish the tree's life force.

After a while, though, you can have too much of a good thing. If too many other males arrive, there will be enough room to build egg chambers but not to accommodate the larvae that will later eat their way out of these chambers and through the bark, forming a starburst of tunnels as they go. The result would be many starving young bark beetles. Once enough males have assembled, therefore, they send out a signal to indicate that the tree is fully occupied to keep further rivals away. The late arrivals aren't left out in the cold, however, as there are usually more spruce trees in the vicinity that can now be attacked. There's a good chance that these too are weakened – at least in central Europe. After all, spruce are not native here, and are therefore always growing in conditions that are really too warm and dry for them. And sometimes bark beetles arrive in such large numbers that they overwhelm even healthy trees.

When whole stands of trees are affected, we talk of 'beetle hot spots'. The reddish crowns of the dying trees stand out from a long way away.

Talking of attracting attention, the beetles' chemical communications have the disadvantage that enemies can 'eavesdrop'. The red-bellied clerid, or ant beetle, for example, which really does look like a large wood ant, hunts down bark beetles by sniffing the scent they give off to talk to one another. And it's not just the adult ant beetles that eat the bark beetles in both their larval and mature phases: their larvae dig in as well. Too much chatting is a disadvantage for bark beetles, just as it is for trees.

As it calls for reinforcements (or turns them away), the little male bark beetle keeps his eye on the prize: finding a mate. It excavates a nuptial chamber under the bark and, using a different scent signal, summons female clientele. Once they arrive, there's first sex and then work – mostly women's work, because there are between one and three females per male. The female beetles construct more tunnels with tiny alcoves for the eggs, which are laid one after the other once construction is complete. All the while, the females continue to mate to ensure they get enough sperm to fertilise thirty to sixty eggs. The bark beetle male doesn't stand idly by, though. Like a true gentleman, he helps by removing the frass.

Later, when the larvae have been left by themselves to hatch, they can eat their fill of the nutritious layers just under the bark and fatten up nicely. In the off-season, you can get a good view of their handiwork on old bark that has fallen off a tree. The farther from the egg chamber, the wider the passages chewed out by the larvae: the increasing width reflects the increasing girth of the young bark beetles. At the end of each passageway, you will find a hole. This is where the beetle emerged from its chrysalis and flew away – but not before forti-fying itself with a final helping of bark. You can see the exit hole clearly if you hold the section of bark up to the light.

Development from egg to beetle takes about ten weeks, which means there can be numerous generations of offspring every year, depending on the weather. Cool, wet summers are hard on spruce engraver beetles, because not only are the trees better able to defend themselves, but it's also easier for fungal infections and other diseases to spread through the insect population. (They dislike long rainy spells as much as we do.)

Fungi, however, are not always a bad thing for beetles. Some species of beetle even need damp wood where these guys have taken up residence. Consider, for example, the conifer ambrosia beetle. It avails itself of timber that is just beginning to dry out slightly. Wood in this state is the perfect place for some species of fungi to settle, because they can't grow in the wet wood of healthy living trees, or in the dry wood of trees that are long dead. The conifer ambrosia beetle leaves nothing to chance. It carries spores of bay bolete on its body and infects the wood with this fungus while it's constructing its tunnels.

The conifer ambrosia beetle goes a layer deeper than the spruce engraver beetle, and makes itself at home in the sapwood, which is the living outer ring of a tree – although not for much longer. It's moister here than further inside the tree, which means the fungi brought along for the ride can spread easily. The beetles construct a system of passageways with short ladder-like side spurs. The fungi begin to grow on all the interior walls, blackening the wood around the passage-ways, and the beetles and larvae use them as a source of food. The combination of blackened wood and holes lowers the value of affected timber – at least as far as forest owners and sawmills are concerned. You can easily distinguish between an attack by these guys and an attack by spruce engraver beetles, because the frass outside the bark is not dark brown but almost white. (It comes exclusively from light wood, after all.)

Holes in the trunk, stains from fungi: it's not surprising that bark beetles are seen as pests. And it's not just that the timber itself is worth less. In warm, dry years, the beetles can multiply so quickly that trees along entire mountain ridges die, as you can see in the Bavarian Forest National Park.

Destruction caused by the mountain pine beetle, however, is on another scale altogether. This beetle lives in pine forests in western North America, where it's particularly partial to lodgepole pines. It behaves much like the spruce engraver beetle, except with them it's the females that lead the attack and summon the males with seductive scents. To shut down a tree's defences – its flow of pitch – the beetle carries a fungus that attacks and paralyses the living layers of bark. That way, not only are the tree's defensive mechanisms shut down, but as the transport of nutrients is blocked it also can't feed itself. The defenceless victim can easily be colonised.

In recent years, there have been increasing reports that these beetles have multiplied so much that they also decimate healthy forests. They've destroyed approximately 55 per cent of all the commercially harvestable pine in British Columbia, and enormous areas have been stripped of all their old trees.[1] You have to wonder why this is happening: usually, a species doesn't destroy its natural habitat. Scientists suggest it has to do with climate change. Higher winter temperatures allow more eggs and larvae to survive, and so the beetles extend their range further north. Warming also weakens trees so that they have less energy to defend themselves against their attackers.

This is certainly part of the problem, but most studies don't mention another important factor: the extensive annihilation of ancient forests, to be replaced with gigantic monocultures that favour such extreme increases in beetle populations. Moreover, rare natural fires, like those started by lightning, are quickly put out, which means that there are many more pines growing in the forest than there used to be. And because these forests are now plantations that put stress on trees, there are many more weak pines, which makes it easier for the mountain pine beetles to proliferate.

This particular beetle is now moving further and further north, and higher up mountain slopes. In other words, it's moving to cooler places – or to be perhaps more accurate, to places where it used to be cooler. Here it encounters species of pine that have never met mountain pine beetles before, and are therefore not very good at defending themselves. The pine beetle's original victim, the lodgepole pine, usually doesn't give in without a fight. When a beetle bores into a lodgepole pine, the tree tries to pump generous quantities of pitch to the wound site. That way it either drowns the beetle, or at the very least flushes it out. Burly beetles, however, wade through the sticky substance and, in so doing, change the chemical composition of the pitch, turning it into an invitation to other beetles to join them and start munching their way into the tree as well.

Once the bark beetle has overcome the first obstacle, it arrives at the living cells of wood. These immediately commit suicide, releasing a potent insect toxin.[2] If there's just a lone beetle, it is killed; however, if it has been joined by colleagues summoned by the chemical call for reinforcement, the beetles will weaken the tree until it is worn down and surrenders.

There are similar wide-ranging forest collapses in Germany. In the Bavarian Forest National Park, large areas of spruce originally planted as commercial forests were put under protection, along with other trees. But once foresters could no longer cut down trees that had been attacked or pump them full of chemicals, the engraver beetle ran riot, just like its cousin in North

America, and with identical results. Entire mountain ridges were covered with dead trees that had been attacked. Hikers were shocked to find nothing but a bleak landscape haunted by cadavers instead of the verdant paradise they had been expecting.

So let's ask ourselves again whether bark beetles really are pests. As far as I'm concerned, the answer is a resounding no. These insects prey on the weak, so they can only damage trees that are already in trouble. The mass reproduction events that allow the beetles to overcome healthy trees only happen when people have changed the natural rules so much that the insects can gain the upper hand. Be it through creating plantations or emitting the pollutants that lead to climate change, ultimately it is us, not the beetles, who are to blame for upsetting the carefully calibrated balance of nature. Instead of blaming the beetles, you could see them as an indication that things are not as they should be. All they are doing, you could argue, is exacerbating a situation that is already out of control, making it all the more urgent that we change course to bring us back closer to the natural order.

The coniferous plantations in central Europe – vulnerable artificial arrangements of non-native species – could gradually be replaced by native deciduous forests. Of course, there are bark beetles adapted to attack these trees too; however, since Beeches, Oaks & Co. are much more at home here than spruce or pine, they usually have no problem repelling bark beetle attacks. So to label bark beetles as pests diverts attention from the root of the problem. Moreover, individual trees that are attacked because they are already weak are a vital source of food for ant beetles, woodpeckers and many other species. Bark beetles, you could say, merely open the door for creatures that live off dead wood. By multiplying in former plantations, they create a temporary paradise for detrivores. And in the ravaged stands of spruce in Germany's national parks, the next generation of trees is already primed and ready to go. That new generation, including many deciduous trees, is standing by to form a solid foundation for the old-style forests of the future, which means that bark beetles are more than just funeral directors: they are midwives too.

It's somewhat easier to understand the process of death and renewal if we look at what's happening to large dead animals. Dead animals? Yes, a dead animal is an ecosystem unto itself, a bit like a small planet in the universe of nature. A planet that's a bit on the smelly side, granted, but one that has been given far too little attention until now.

7

The Funeral Feast

So FAR, WE'VE PASSED OVER a particularly tasty treat for many species: the carcasses of large mammals. Fascinating things happen around these dead bodies. Do you find that disgusting? That's understandable; however, strictly speaking, we're surrounded by the dead bodies of animals all the time, and unless we're vegetarians, we interact with them (albeit briefly) almost daily on our plates. The main difference between our meals and the many dead boar and deer out in the wild is that on our dinner table the process of decomposition has barely started, which allows us to enjoy our meals safely.

Many animals tolerate or even require various stages of putrefaction in their food. They are perfectly happy devouring servings of meat that to us would stink to high heaven. And there are a lot of servings to be consumed. Every year in central Europe alone, millions of roe deer, red deer and wild boar die a violent death. And although in Germany, for instance, a lot of wild game is shot – about 1.8 million of the three aforementioned species, according to the German Hunting Association – many more die a natural death. What happens to their bodies? Well, they decay, I hear you say. In other words, they rot and eventually, after smelling awful for a while, they become humus. But who facilitates this process?

Let's start with the larger facilitators. Bears have extremely sensitive noses and can smell a side of meat from many miles away. Together with other large predators, such as wolves, they can consume most of the meat off a dead animal within a few days. What they can't eat, they bury, so they have a hidden supply of food. Birds are early responders, as well. Whereas vultures circle over fresh carcasses in the African savanna, noisily laying claim to them, in northern latitudes ravens take their place. Ravens are the vultures of the North, and they too patrol their territory from above to see where a deer or wild boar might have met its end.

Dead animals are often the cause of fights, and when brown bears turn up wolves lose out. Then it's best for the pack to head for the hills, particularly if they have pups, which a bruin could easily scarf down as a snack. Ravens offer valuable assistance here: they spot bears from afar and help wolves by alerting the pack to approaching danger. In return, wolves allow ravens to help themselves to a share of the booty – something the birds wouldn't be able to do without the wolves' permission. Wolves would have no difficulty making a meal of ravens, but they teach their offspring that these birds are their friends. Wolf cubs have been observed playing with ravens; they remember their smell and come to regard them as members of their community.

Wolves and ravens might live peaceably with one another, but other species fight over food resources. Apart from the black birds, there are other feathered parties, such as bald eagles and kites, that would love to haul away a portion of the booty. With all the commotion and clamour as animals wait their turn, the ground around the carcass gets torn up. The plants get moved about, because seeds that would otherwise have been stifled in the matted grass now have their moment in the sun. Things also change even for vegetation that is not disturbed. Rotting flesh serves as fertiliser – for the plants, deer carcasses are simply overgrown salmon. The more robust growth and greener colour of the grass and non-woody plants for about a metre around the carcass are evidence of the nutrient boost.[1]

And what happens to all the bones? Even if the flesh gets eaten or rots in the way I've said, there should be large amounts of bones lying around in field and forest, bleaching in the sun. But on my daily rounds as a forester I have never come across a dead animal's final resting place, and only very occasionally do I find a skull.

There are two forces at work here. Sick or weak animals separate themselves from others of their kind to hide in the undergrowth, or on hot summer days wander near or even into small streams to cool any wounds they might have. Here they wait for death. It makes sense, because this way they don't endanger their kin – for weak animals attract the attention of predators. Also, in a secluded spot there's no one to disturb them in their final hours. Usually it's the smell of decay that leads us to dead animals in places like this; once the flesh has rotted away, the bones lie quietly hidden from sight under bushes. Because bones basically don't break down, and because every now and then animals surely must die away from the protective cover of vegetation, over time you'd expect to find bones scattered all over the place. But that's not the case, because there are plenty of takers for the dead animals' final remains.

Mice, for example. They seem to love bones, and gnaw away at them until there's nothing left. Calcium and other minerals are mostly what they're after: bones are for mice what salt licks are for cattle (or salted pretzels for us). If the bones are still fresh, bears eagerly crack them open for the fatty marrow inside – a delicacy that no one will fight them for, not even wolves. Although some dogs like to chew on bones, wolves clearly don't think much of this tedious task, but it is an important one for other species. Just how important becomes clear everywhere bears have been eradicated, including in Germany, because it is only when the hard outer coating has been cracked by these large mammals that daintier creatures get their turn.

Take the bone skipper, which had disappeared without trace until it was recently rediscovered.[2] With its tiny orangey-red head this bizarre insect looks like a creature from a fantasy world, and it doesn't behave like other flies, either. The bone skipper likes it nice and cold. It's out and about mostly on winter nights, on the lookout for dead animals and cracked bones, which is where it feasts and lays its eggs. By the nineteenth century in central Europe, however, thanks to stricter rules on hygiene, there were no longer any carcasses out in the open. At the same time, bears were driven out, and so things became grim for the bone skippers; in 1840, they were declared extinct. In 2009, however, the Spanish photographer Juli Verdú took a picture of what he thought must be a fly that had arrived from the tropics. Researchers at the University of Madrid recognised it as the long-lost insect, which could then be crossed off the list of extinct animals.[3]

I spoke earlier of ravens as the vultures of the North, but we should also mention vultures themselves. Griffon vultures regularly fly over Germany searching for dead animals. On the website Club300, amateur ornithologists report sightings of these unusual visitors every year.[4] If there were something for them to eat, several of them might once again make their home in Germany, but as it is, all they do is make flying visits that pass unnoticed by most of us. Like bone skippers, Griffon vultures have been declared locally extinct in many places in the world.

Until now, we've been focusing on the carcasses of large animals. These are usually fastidiously disposed of, but below a certain size, this clean-up no longer happens. The remains of small mammals vastly outnumber those of large ones. Take mice, for example. At any given time up to 100,000 of these little rodents scurry around every square kilometre, living on average just four and a half months. Young mice are sexually mature by two weeks, and after another two weeks females give birth to about ten babies.

Let's assume that, during one growing season, there are five generations of ten mice each for every mouse pair. In particularly fruitful years, that would mean the population of 100,000 animals (or 50,000 pairs) per square kilometre increases to 2.5 million mice scampering around – not all at the same time, of course, because as the season progressed most of them would have died from disease or been eaten. Over the course of the season, then, there could be up to 6.5 million dead mice lying around. If each weighs an average of 30 grams, the total weight of these dead bodies would be 75 tonnes, the same as about 3,000 roe deer. That's way too much mouse meat to be carried off by buzzards, foxes or cats, so it leaves plenty for others to exploit.

One of these is a pretty black-and-orange-striped beetle appropriately called the sexton or burying beetle. I encounter it frequently on my walks through the forest – it's so striking it's hard to miss. Even though the adults hunt insects, they can't resist the delectable scent of fresh carrion. For sexton beetles, a mouse carcass is attractive not only as a hearty meal but also as a good place for their offspring to get their start in life. It's often the males who get the first opportunity. They triumphantly raise their rear ends and discharge a scent message to attract females. Their goal is clear: sex. Rivals, however, also get the message and fly on over. Fierce struggles ensue, and the losing beetle has to beat a hasty retreat. When a female turns up, the work begins.

The beetles dig tirelessly underneath the dead mouse, dragging it down by its fur. In the process, a lot of the fur gets bitten off and the carcass gets coated with generous quantities of saliva. That doesn't sound very appetising, but it makes the mouse more slippery. And so the dead animal gradually sinks further down into the soil, until eventually it disappears completely, safely beyond the reach of other carrion eaters.

The beetles take frequent breaks to have sex. After all the work is over, the mouse doesn't look much like a mouse any more. All that pulling and pushing has transformed the carcass into an elongated pellet. The female now lays her eggs alongside it. Unlike many other insect parents, sexton beetles stick around after their larvae hatch. The youngsters' mouthparts are not strong enough to chew meat, and so the mother feeds her little ones, which lift their heads and beg for food like baby birds in a nest.

As researchers at Ulm University discovered, something else happens to the beetle mother: she loses interest in mating. Not only that, but even if the male were to get lucky, it wouldn't do any good, as his beloved is now completely infertile – at least as long as she still has her full complement of babies. As soon as a couple of the little ones go missing, though – perhaps because they died or were eaten by some animal – her desire for sex returns. The male immediately gets wind of the change and goes berserk. The scientists counted up to 300 copulations – more than after the male had initially laid claim to the carcass. The female quickly lays new eggs to replace her loss. If, in this flurry of activity, she ends up with too many babies, she soon fixes things by killing the extras.[5]

If neither bears nor wolves take care of a carcass (or, in the case of smaller carrion, sexton beetles), then even smaller creatures take over. Heading this squad of cleaner-uppers are blowflies. In Germany alone there are more than forty species that are magically attracted to the smell of dead bodies. The meat doesn't even have to be so far gone that it stinks: in fact, these insects prefer to descend on fresh food. So if you leave a portion of barbecued meat on your plate in summer, it often takes only a few minutes for the first flies to appear.

I found out for myself just how fresh these iridescent blue flies like their meat. On a hot summer day some years ago, I came across a roe deer that had lain down in the undergrowth. It was badly injured, and had a large wound on its hindquarters. There were already hundreds of fat white maggots crawling around in it – the children of blowflies. With a heavy heart, I put the deer out of its misery. Some species, such as toadflies, even attack perfectly healthy animals. They lay their eggs on the skin of toads and, when the larvae hatch, crawl up into the toad's nostrils, where they begin to eat their host's head from the inside out. The toad briefly shuffles around zombie-like before finally giving up the ghost.

Usually, blowflies are the first guests to arrive at a fresh carcass. Hundreds of flies lay thousands of eggs, preferably in exposed places like the eyes. The fast-developing larvae spread from there to the whole carcass and cover it so completely that other insects have little chance of finding an empty spot to lay their eggs. The bone skippers are the last to appear, content as they are with the leftovers, which is to say: the bones.

There's a simple way – in Germany, anyway – to support bone skippers and numerous other species that depend on very large carcasses. We could let dead deer and boar lie, at least in national parks. When people hunt in these parks, the carcasses of wild animals are usually taken away by foresters. But in national parks, natural processes are supposed to be able to play out, and the carcasses of animals are and should be allowed to be part of this cycle.

We won't be able to see the red-headed bone skipper for ourselves, because it's mostly active on cold nights. Even so, it's good to know that this ecosystem, with its sometimes bizarre-looking creatures, has another chance at survival.

Talking of the night, there are other representatives of the insect kingdom that also like the dark, and even light small lamps of their own. These lamps light the way to love, intrigue and sometimes even gruesome death.

8

Bring Up the Lights!

<hr />

IN NATURE, LIGHT is of the utmost importance. Ultimately, almost every creature on the planet lives off processed solar energy. Photosynthesis produces sugar, which fuels plant life and therefore, indirectly, human and animal life too. In the natural world there's a struggle for every ray of sunlight, for every smidgen of energy. Trees demonstrate this particularly well. The only reason they grow as tall as they do is in order to rise above the other plants and bushes that compete with them for light.

Developing mighty trunks and crowns takes a great deal of energy. A mature beech, for example, contains up to 13 tonnes of wood, which, if burned, would release about 42 million kilocalories of energy. To put this in perspective, a person, depending on how active they are, burns between 2,000 and 3,000 calories of food a day. (Food calories are actually kilocalories, even though we refer to them simply as calories.) This means that a mature beech stores enough solar energy to feed a person for forty years – if the human gut were able to digest wood. It's no surprise that it takes decades to produce this much wood, and that's why trees have to live to be so old. A forest ecosystem, then, is basically an enormous storehouse of energy.

So far, so good, but light is also important for completely different reasons. Its energised waves stimulate the retina at the back of the eye, where they are transformed into information. Most animals have developed their vision to interpret light, which means, of course, that some light needs to be available. Apart from the fact that the enormous crowns of trees block up to 97 per cent of the light in a forest, there's yet another problem for animals that need light waves in order to see. Half the time – at night – there's precious little light around. Only the faint glow of the stars, augmented by the brighter light of the moon, offers some relief from the darkness. But when it's cloudy, as it so often is, it can get pitch-dark out there. So why not make a virtue out of necessity?

Although the title of this chapter is 'Bring Up the Lights!', for some plants and animals 'Turn Down the Lights!' would be a better call. Plants and animals are nocturnal for many different reasons. Some flowers bloom only when it's dark, because they want to avoid competition. During the day, a multitude of non-woody plants, bushes and trees, all vying for the attention of pollinating insects, go to great lengths to stand out from the crowd. Let's consider honeybees, which are important pollinators. They're limited in how many flowers they can visit, and if the floral host is too large, then many flowers miss out on pollination and don't form seeds. To avoid that fate, plants make use of the full range of colours in nature's palette. In addition, they send out sweetly scented invitations. What smells good to us also smells good to insects, because sweet smells indicate where delicious nectar is to be found.

Some plants opt out of the colourful daytime chorus of visual and olfactory communication, and shift their bloom time to the hours of darkness. Their names – evening primrose or moonflower – often point to their nonconformity. After the sun sets, most other flowers shut down, so you could say that the competition goes to sleep. Now the insects can focus all their attention on the few remaining plants offering nectar. It's too bad that the honeybees, like most of the flowers, also stop work and take a break. They returned to the hive a while ago to spend the night processing their haul and preserving it as honey.

But there are insects that work the night shift. Take moths, for example. I must confess that, even though I'm an animal lover, moths are not high on my list; but in my defence there's a story here. Years ago, when we came back from a family holiday in Sweden and, after unpacking the car, finally collapsed onto the sofa, we noticed that there were tiny moths flying all around us. A nasty feeling came over me. I lifted a corner of our woollen rug. Horror of horrors! Thousands of larvae were wriggling about in the wool, and a blizzard of disturbed moths flew up and fluttered around the living room. We quickly rolled up the rug and banished it to the garage. The incident made me feel queasy around moths, and a flicker of that queasiness still returns any time I come into contact with wool.

Moths and butterflies belong to the same order (Lepidoptera), but there are clear differences between them. One is that butterflies fly during the day, and most moths fly at night. Another is that butterflies are colourful, whereas moths tend to be rather drab, but there is a good reason for this. Butterflies use their colourful patterns to communicate information to other butterflies and enemies, whereas moths have a completely different strategy. If they are to survive, they need to remain as unobtrusive as possible and blend in with their surroundings as best they can, because these little winged creatures spend the day somewhere on the bark of a tree, where they have to avoid the attention of birds that might eat them.

But at night birds go to sleep, which puts moths at a distinct advantage when they visit the sweet chalices of night-blooming plants. How lucky for moths that most birds agree with the plants: the hours of darkness are not to their liking and they avoid them. But because this interplay between species has been going on for millions of years, it should come as no surprise that predators are standing by to exploit the situation.

These predators are bats, which in the warmer months of the year hunt moths. And because light is in short supply at night, bats use ultrasound to locate their prey. I think it's entirely possible that bats use their calls and the sound waves reflected back from objects to help them construct images inside their heads – in other words, they 'see' objects using sound. Scientists assume that, thanks to the echoes bats receive, these nocturnal hunters have a very clear idea of who or what they are dealing with. A leaf falling from a tree creates a different ultrasound pattern from the flap of a moth's wing. Bats can detect wires only 0.05 millimetres thick, and it's possible that they 'see' their surroundings in much more detail than we do during daylight hours with our eyes.[1] After all, when we see things, we're doing nothing more than interpreting waves reflected off objects. The only difference is that we're adapted to light instead of sound – that, and the fact that bats have to shout all the time if they want to see something.

This shouting is not long and drawn out, as it is when we want to elicit an echo while we're out hiking in the mountains, for example. Unlike us, bats make a series of short, rapid calls – about 100 per second. And with bats, it's all about volume: the sounds they make can be as loud as 130 decibels, which would hurt our ears if we could hear them (but thankfully ultrasounds are out of our range). Unlike lower frequency sounds, ultrasounds are quickly swallowed up in the air, and after they've travelled about 100 metres, there's not much left to hear. Even so, on summer nights it can get quite noisy out there in the forests and fields – in the upper frequencies, at least.

To camouflage yourself from the reflection of light waves, or, to put it more simply, from being seen, all you need are colours and patterns that blend in with your surroundings. The same is true if you want to camouflage yourself from sound waves. In this case, blending in with the surroundings means that, as a moth, you reflect back as little sound as possible. You can test out how this works the next time you hike in the mountains.

Your calls to elicit echoes are reflected particularly clearly if the slopes around you are not covered in trees. If they're forested, you usually won't get an 'answer', because the trees' trunks and crowns swallow sound. To exploit this effect for their own ends, moths grow their own mini-forest. Their bodies look furry, and these 'hairs' ensure that the sound waves are not reflected back crisply and clearly, but are deflected in different directions so the bats can't get a clear picture. Unfortunately, this deflection has its limits, so these insects need more tricks to increase their chances of survival.

There's a regular arms race going on between moths and bats, and at least some of the moths are catching up. Over time, some moths have evolved to hear sounds at extremely high frequencies. The highest frequencies bats use when they hunt are around 212 kilohertz. In contrast, human hearing breaks down at frequencies higher than 20 kilohertz.

Although moths can hear higher frequencies than we can, they cannot match the frequencies of bats. The result is that the moths can't hear the bats coming (bat wings make hardly any noise at all), and when the predators attack they are taken completely by surprise. But that's not the case for every species, as Hannah Moir and a team of researchers at the University of Leeds discovered. The greater wax moth can hear sounds in the range of 300 kilohertz – the highest hearing score in the animal kingdom – even though its ear has a really simple design: it is made up of a membrane with just four receptor cells. (In comparison, there are 20,000 so-called hair cells that vibrate in a human ear, in addition to other structures that convert sound into neural signals.)

As Moir and her colleagues report, the moth's hearing is way more sensitive than it needs to be. If bats don't produce sounds much higher than 200 kilohertz, why do moths need to be able to hear higher frequencies? And as bats are unlikely to upgrade their calls, sounds in frequencies higher than the ones they are currently using are quickly muffled by the air around them, and are therefore not helpful when the bats are trying to create echoes.

Why, then, have greater wax moths developed this extraordinary ability? The researchers speculate that the moths probably have something completely different in mind. They, too, communicate at high frequencies, for example to find a mate. Their closely spaced courtship calls are within the range of the bats' more widely spaced echolocation calls. The simple construction of their ear means that greater wax moths can distinguish closely spaced signals better and faster – six times faster – than other moths. (Some individuals, it turns out, have more sensitive hearing than others.) Consequently, the moths can flirt in peace, because they can hear the echolocation calls of their greatest enemy loud and clear over their own calls, and can make themselves scarce if they need to.[2]

The greater wax moth is not the only species to have armed itself against bats. Some moths interfere with bat echolocation systems by producing decoy calls – clicks in the ultrasound frequency that confuse approaching bats. Effectively, this causes the moths to disappear amid static on the bats' radar. The great tiger moth, a member of the subfamily Arctiinae, many of which have hairy caterpillars commonly known as woolly bears, produces such a fearful din that alarmed bats call off the chase.

But how do moths escape once they've heard their enemy? Bats fly much faster than moths, and they're also more manoeuvrable. This means there is just one basic defensive response when danger is closing in. As soon as moths capable of hearing ultrasonic calls register the sounds of a search, they fall to the ground in terror: the bats have little chance of tracing their prey in the grass. Despite the moths' defensive tactics, however, at night the predators eat their fill. There are always careless moths to be caught, and mosquitoes too. Bats catch up to half their own body weight in insects a night – and on a night of dining on nothing but mosquitoes, that would be about 4,000 mosquitoes per bat.

Hunters and prey coexist in a delicate balance that gives each a chance to survive. But artificial light can upset the scales. At night there is, of course, just one relevant source of natural light: the moon. When it shines, animals use it to orient themselves. It serves as a kind of compass. When moths are out at night and want to fly in a straight line, they are careful to keep the celestial body at a set angle to their flight path. That works wonderfully well – until an artificial light crosses the tiny flier's path.

The insect assumes this light must be the moon. Confused, it tries to fly so that the moon is always on the correct side – for example, to its left. With the moon, that's not a problem, because it's infinitely far away. With the light, which is close by, the insect quickly flies past. The source of the light is now behind it. It keeps correcting its course, and these corrections lead it to fly in an ever-diminishing circle. Eventually the moth crashes right into the light itself. It keeps trying to get away – but every new attempt at escape ends in failure.

Some moths die of exhaustion; others meet their end more quickly. Over time, many bats have specialised in patrolling street lamps. They can eat their fill of insects more quickly here, because all they have to do is check out one lamp and then the next to see if yet another moth has become confused by the artificial moons. Even windows lit up at night can set the stage for similar small dramas, as my wife and I have observed. As we lounge on the sofa watching a film, moths gather at our living-room window. Every now and again, shadowy bats flit by – until finally all the moths have disappeared.

There's a whole host of other insects that become confused in a similar way by artificial light. Like moths, they're magically attracted to garden lights that are supposedly lighting up their surroundings in an environmentally friendly manner. They're usually powered by a solar cell, which is ecologically responsible and means there's no issue with leaving the light on all night long. This delights the large number of spiders that benefit by spinning their webs here. In the long run, the tiny ecosystem around the light changes, because some species disappear entirely (into the spiders' stomachs). If it were just one light, that wouldn't matter so much, but it's different when there are thousands upon thousands of them, as there are in urban areas.

There have, however, been additional sources of light in the landscape since long before people started lighting things up. On warm summer nights, thousands of tiny greenish lights glimmer along the edges of forests and shrubby areas. These are fireflies (sometimes called lightning bugs), which strut their stuff when it gets dark. The brightness of their light may be a thousand times less than that of a candle flame, but the efficiency with which they turn energy into light is exceptional. Using our latest technological advances, we can convert 85 per cent of energy into light; fireflies can convert 95 per cent. They need to be frugal, because as adults they don't eat anything – at least not in most cases (there are gruesome exceptions, but more on them later).

Fireflies really should glow red, because the purpose of their nocturnal light show is love. In the most common species in Germany, *Lamprohiza splendidula*, the females light their lamps on the ground. Fireflies are therefore also known as glow-worms, even though the light you see at night comes from the adult beetles. The females have stunted stubs instead of wings and cannot fly, so, with their pale yellow abdomens dotted with tiny lights, they really do look like luminescent worms.

These earthbound females don't turn on their lights until they spy a male above. Males can fly, and they scour the neighbourhood in search of a mate. Their last two body segments are protected by a transparent chitin casing that allows them to shine their light downwards. That way they don't reveal their presence to enemies flying above them, while at the same time signalling 'look what a great guy I am' to the females below.

When one of the females gets the message, she turns on her light as well, inviting the Casanova to land, which he does right away. There's mating and then egg-laying. The larvae that hatch eat a great deal. They're partial to snails, and will tackle specimens up to fifteen times their own weight.[3] They kill the snails with a single bite, and then eat them at their leisure. The firefly larvae keep eating until they're ready to burst, and when their stomachs are full, it's time to take a nap. How long they pause to digest depends on the size of their meal, but their post-prandial nap can last for days.

Depending on the species, it can take about three years until the offspring develop into sexually mature beetles. Given how long they spend in their larval stage, the common name glow-worm seems entirely appropriate. The adult luminescent beetles live for only a few days: the male dies soon after mating, and the female gives up the ghost right after she has laid her eggs. And so their glow is literally the final flicker of a life that ends on an ecstatic high. At least it does when all goes according to plan. Unfortunately, there are always disruptive elements lurking out there in nature.

The glow-worm's peaceful illumination with love in mind is abused by others for their own ends. In New Zealand and Australia, there are gnats in the genus *Arachnocampa* whose larvae also light up. These antipodean glow-worms live in caves and deep in the rainforest, where they gather in groups high up on the ceiling or in the canopy. The larvae need a dark, humid place where the air is still – conditions that only caves and dense forests can fully provide. They then spin sticky threads covered in tiny drops of moisture that light up.[*] The effect is magical, and caves where the larvae live have become popular tourist attractions. But wealthy tourists aren't the only ones attracted to the lights. Insects also visit, probably because they confuse the twinkling droplets with stars in the night sky. Thinking they are flying out in the open, they get caught in the sticky threads and end up in the stomachs of the hungry gnat larvae. The hungrier the larvae are, researchers have discovered, the more brightly they glow.

A North American beetle in the genus *Photuris* employs an even more deceitful tactic. Fireflies have developed a variety of ways of using light to draw attention to themselves. There are different species, after all, and if all they did was simply light up, when it came time to find a mate they could easily get confused. So there's a kind of Morse code out there – flashes with a particular rhythm and frequency that attract beetles of the same species. Human Morse code would be too primitive for the beetles: on/off and long/short just wouldn't contain enough information. Using up to forty flashes a second with varying degrees of brightness, the beetles are capable of a much wider range of signals.[5] And so cheery winks are a way to find the love of your short life – unless you belong to the genus *Photuris*.

Females in this genus imitate the light signals of a different species to lure their males, which flock eagerly to them. When the males land, instead of an amorous adventure, they find the eager mandibles of *Photuris* females. They need the males not only for their calories but also for the toxins in their bodies. These then protect the females from being eaten by spiders, which also notice their light signals and, in the absence of toxins, would be happy to accept the illuminated invitation to dinner.[6]

It's not just insects that use light as an attractant. The deep-sea anglerfish possesses, as its name implies, a fishing rod. This sits on top of its head, dangling a bioluminescent lure out in front of its mouth – a mouth that bristles with needle-thin, dagger-sharp teeth. The light source attracts other fish like a magic wand, and you can imagine how their visits end.

Humans get similar results when they use lights to fish. Fishermen in Japan, for example, use this technique on a massive scale. Light is tremendously attractive whether you're on land or in the water. And this brings us back to the problem of humans lighting up the night. It is startling when you look at a satellite map of the earth at night and see how much of its surface is lit up with artificial light. You can easily gauge for yourself how severe the impact of artificial light is where you live by stepping outside your front door at night. Can you see the Milky Way on a clear night? If you don't even know what the Milky Way looks like, then there are certainly too many artificial lights in your neighbourhood: if conditions are favourable there's no way you can miss this impressive band of light.

Visibility is further diminished by air pollution, which diffuses light particles, so the number of stars visible to the naked eye can decrease from about 3,000 to fewer than fifty. And the delicate light signals from fireflies are a little bit like weak stars. The more artificial light there is in this world, the more confusion there is in the animal kingdom of the kind I've been describing, and the less successful those species that produce light themselves.

The confusion can be fatal. Freshly hatched sea turtles orient themselves towards the glittering waves of the ocean that are lit up by the full moon. As soon as they've scrabbled their way out of their sandy hiding places, they set out in this direction as quickly as they can to escape hungry predators. Problems arise if the beach is close to a brightly illuminated sea promenade or a strip of hotels. Then the tiny turtles set out towards the artificial light sources, getting further and further away from the safety of the water. It's no surprise that the next day many of them fall victim to gulls or die of exhaustion.

Even weather phenomena get turned upside down thanks to the glare of electric lights. Clear nights used to be particularly bright, which makes sense because the moon and the stars can shine down uninterrupted. Once our eyes have had a few minutes to adjust to the dark, we can go for a night-time walk without any problem. Today, that's even possible on cloudy nights – which used to mean complete darkness – because clouds reflect urban lighting way out into the surrounding areas, inadvertently lighting up the sky in a way that does neither people nor animals any good. After all, who likes to sleep with the light on?

And yes, artificial lighting has negative consequences for people too. The internal clock that ticks inside us is regulated by light. Its blue wavelengths are particularly important for us, because they determine whether we feel wide awake or tired. Our eyes contain the photo-pigment melanopsin, and when it's hit by blue light, it signals to our brain that it's daytime. Usually, the system works really well. In the evening, at sunset, the light spectrum shifts to red and we automatically feel tired.

Problems arise when we watch television at night instead of going to bed, because the flickering images on the screen contain a great deal of blue light. It's no wonder so many people suffer from sleep disorders. Staring at the screen primes the cells in our body to be ready for action rather than sleep. The makers of smartphones are trying to get a handle on the problem by adjusting the screen colours after a certain hour so that users feel sleepy as they surf and chat.

What about animals? How can we help creatures that can't escape all this light? You can give your fellow creatures at least some relief simply by closing your blinds or drawing your curtains at night, thereby blocking a widespread source of light. And you don't need to keep your outside lights on all night, either. Along the driveway leading up to our forest lodge we have motion-sensor lights which come on briefly only when they're needed.

The greatest source of night-time illumination, however, is street lamps. These days, most radiate an orange-red light that reflects particularly well off clouds, making the problem of lighting up the night sky even worse. I was pretty excited for a while when the old white fluorescent tubes were replaced by these modern, energy-efficient sodium vapour lamps. But then I noticed that the undersides of clouds were glowing increasingly red, and there were nights when the radiance in the clouds drew my attention to the city of Bonn a full 40 kilometres away. I used to attribute the increasing brightness of the night sky more to the ever-growing sprawl of the city. Not any more. Now there's a trend to switch to LED bulbs, which use even less energy. And if these street lamps were more focused – in other words angled to only shine downwards (which is where the light is needed) – and if they were also turned off after midnight, these would be big steps in the right direction.

Although there's still much to be done to protect the environment at night, when the sun lights up the day there are signs of heartening ecological improvements. Here, in autumn, you see impressive bird flocks pass by in formation on their way south – with unexpected consequences for the production of Spain's famous Iberian ham.

9

Sabotaging Ham Production

I LOOK FORWARD TO AUTUMN every year, or, to put it more precisely, I look forward to the cranes. You can hear the trumpet-like calls of the migrating flocks from many miles away. And after all these years of keeping an ear out for them, I can pick up their distant calls even if my living-room windows are closed. Over recent decades, thanks to better environmental policies like the reclamation of wetlands, the number of birds has risen to such an extent that Eurasian cranes are no longer listed as endangered. Day after day, one formation after another flies over our forest lodge, and sometimes the birds fly so low that I can hear the sweep of their wings.

What drives birds to fly to distant lands as the seasons change? How do they find their way? Bird migration is a worldwide phenomenon undertaken by about 50 billion birds. These mass aerial movements are happening all the time, because somewhere in the world there's a change from summer to autumn, from winter to spring, or from the rainy season to the dry season. And as seasons change, so do food resources.

Once the cold sets in here in the Eifel mountains, insects hibernate, dozing deep underground or under the bark of mighty trees. Some species even make themselves comfortable in the relative warmth of the hills constructed by red wood ants. In their chosen hidey-holes, the insects are mostly out of the reach of birds, as are most other small animals that birds prey on, and so many feathered species set out for warmer, more productive climes.

Most researchers believe that the urge to fly to other places as the seasons change lies in the birds' genes. To me, this makes the birds sound like some kind of biological automatons operating in response to a pre-programmed code, incapable of making their own decisions about where and when to fly. But apparently they *do* make decisions, as the Estonian scientist Kalev Sepp and his colleague Aivar Leito discovered.

Since 1999, Sepp and Leito have fitted numerous cranes from their homeland with radio collars so they can track their migration routes. To the scientists' surprise, they discovered that over the years the birds switched between three different routes – which demonstrates clearly that a route hasn't been genetically fixed. It also seems to rule out the cranes learning the route from older birds – until then another theory entertained by scientists. Somehow, Sepp concludes, the birds must get together and discuss where they have the best chance of finding good breeding sites and food.[1] Which brings me to the subject of this chapter.

Cranes, with their assignations and meetings in particular places, really do sabotage the production of Iberian ham. Not intentionally, of course. The birds have not the slightest interest in pigs. But they are well aware that a special treat awaits them in Spain and Portugal: acorns, and particularly acorns from the holm oak woodlands in the Extremadura region in Spain, where they know they can find copious quantities of them. No wonder the cranes that fly over our forest lodge have decided that this paradise is the place where they will spend the winter. Here they can build up their strength and survive the cold season well nourished. However, other inhabitants of the Extremadura also prize these treasures, and these are the local farmers who rely on the acorns to fatten their pigs.

What we're talking about here are the famous Iberian pigs from which *jamón ibérico de Bellota*, or acorn-fed pure Iberian ham, is made. Most of the pigs are raised in an environmentally friendly manner: they spend time roaming through holm oak forests, and a good portion of their diet is the non-woody plants they find there and, above all, acorns. And that's how it used to be in central Europe too. Pigs were driven into the forests in the autumn to fatten up on acorns and beech nuts. (In those days, animal fat was still prized.) The term 'mast years' dates back to these times: these are years when there is massive production of acorns and beech nuts, which tends to happen every three to five years.

But back to the Extremadura. The holm oaks were once an important part of the ancient forests that grew here. Over the course of the many thousands of years of human history, most of the forests on the Iberian Peninsula have been cleared. Different species of trees have been planted, and the character of the landscape has changed. And so today, in addition to conifers, there are more and more eucalyptus plantations. Eucalyptus grows rapidly, much more rapidly than the ancestral oaks, and as such is a good tree to grow if you want to produce more timber.

These changes have been catastrophic for native ecosystems. Eucalyptus plantations, in particular, are considered by environmentalists to be green deserts. The trees' essential oils (which taste so refreshing in throat lozenges) are responsible for an explosion in the number of forest fires. Today, we associate southern Europe with forest fires, but that never used to be the case. Left to their own devices, natural deciduous forests do not burn, so in these latitudes fire was not part of the natural ecosystem.

These changes make the remaining holm oaks that much more important, even if today they no longer grow naturally, but need a helping hand from farmers to get started. The driving force for farmers is not only the need for timber but also the production of acorns for their pigs. And this is where the cranes enter the picture. It's not a problem for the farmers if the birds help themselves to some of the nuts. The problem is the sheer number of birds. According to the World Wide Fund for Nature, in the 1960s there were only about 600 breeding crane pairs left in Germany. Since then, the number has risen to more than 8,000. Over their whole range, which includes the northern parts of Europe and a large part of northern Asia, current estimates peg the population of Eurasian cranes at about 300,000 individuals. And when autumn comes, more and more of them are heading for Spain.

Obviously, more cranes means less food for the pigs, which affects ham production. It's a dilemma. Keeping pigs makes people preserve oak forests, which in turn provide important winter food for cranes. If pig farming loses its lustre, at least part of the motivation for preserving oak forests is also lost.

Is there a way out of this dilemma? I think there is, and the solution is simple: more deciduous trees in Spain and Portugal would benefit all parties. It's true that oaks don't grow as rapidly as eucalyptus or pines, and they are not as easy to maintain on an industrial scale. However, they produce timber that is sought after, and they also provide feed that farmers want for their pigs, which other plantation trees can't offer. Moreover, if more oaks were grown, the danger of forest fires would be reduced considerably, and the ecosystem would become attractive to other species again. (We haven't even touched on squirrels, jays and the thousands of other animals and plants that depend on oak forests.)

Of course, in a democracy you can't just issue a decree to increase the size of forests, but I believe subsidies (which I don't usually support) would get us there.

If the factory farming of animals profits from government incentives I can't see why it should be difficult to do something to promote the peaceful coexistence of pig farmers and cranes. After all, it's not the birds that are overtaxing the ecosystem: what makes the problem so acute is that so little oak forest remains. But if one day there were many more holm oaks, wouldn't that mean that the crane population would go into overdrive? No, it wouldn't. The number of cranes depends mainly on the size of suitable breeding sites. And, unfortunately, wetlands continue to shrink in Europe, which means that the increase in numbers will eventually level off.

If we were all to dial down our demands a bit, there would be enough space for our fellow creatures. In this respect, the crane is a good ambassador for the environment, and one that will, I hope, be around for a long time and in great numbers, with its noisy flight and trumpeting calls to remind us of the early days of the conservation movement.

But what are we to do until the oak forests expand? Couldn't we just feed the cranes while we're waiting for them to grow? That brings up a basic question about supporting our feathered friends, and this question has little to do with science and more to do with our emotions. Shouldn't we feel sorry for birds in winter? Those that don't fly to warm southern climes sit freezing on the branches of bushes and trees, puffed up like fat feathery balls while we watch them through the windows of our centrally heated homes. As they, like us, are warm-blooded, they have to keep their bodies warm. And for birds this means maintaining body temperatures between 38 and 42 °C – which is substantially warmer than ours.

Luckily, nature has equipped birds with an outfit that helps them keep warm: a snug feather coat. There's a reason why we stuff our winter jackets with down: it's an excellent insulator. When birds fluff out their feathers, they trap a thick cushion of air, and when they puff out like this, their spherical shape shrinks their body surface in comparison with their overall volume. Birds also have a cooling mechanism for their legs: the blood flowing into their feet loses warmth to the blood being pumped back up from their feet, lowering the temperature of these naked extremities to almost 0 °C. That's why waterfowl can paddle around in icy ponds without feeling pain.

Despite these adaptations, the smaller the animal, the larger the surface of its body relative to its volume. So, per kilogram of weight, a bear has far less skin than a small bird; therefore, per kilogram, it also loses far less warmth. This means that very small birds – the goldcrest, for example, which weighs only about 5 grams – have formidable problems producing enough energy in the form of heat. Incidentally, the delicate song of the goldcrest is a good way to check your hearing. Its frequency is so high that many people over the age of fifty can't hear it at all. (These days I can barely hear it myself.) Unfortunately, the bird's voice doesn't help it keep warm, and it must replace the energy it's constantly losing through its skin and feathers, or the little singer will soon freeze to death, which basically means it has to eat all the time.

While bears sleep in the comfort of their winter dens, Chickadees, Robin Redbreasts & Co. are constantly searching for calories. But there are often not enough to go round. Beetles and flies have retreated deep under the leaf litter on the forest floor or are dozing in the dead wood of fallen trees. And berries and seedpods are either buried under snow or have already been eaten. No wonder many birds starve, most in their first year of life. The average life expectancy for European robins is little more than twelve months, even though the birds are capable of living for four years or more – if they have enough to eat.

When you see a small feather fluffball sitting there freezing in your garden, don't you feel sorry for it and feel you must do all you can to help? For the first fifteen years we lived at the forest lodge in Hümmel I was very strict. Feeding birds meant interfering with nature by changing their food situation in ways that were not natural. When you install a bird feeder and provide grain and fat, you promote the population of specific species of birds. Many of the young then survive the winter, and the next spring these species are particularly populous – at the expense of others that perhaps did not come to the feeder. There's also the fact that in nature, reproduction rates are perfectly calibrated to winter losses. Species that lose more individuals over the winter simply lay more eggs and breed more than once a season.

For years, therefore, despite pleas from my children, I refused to interfere. In retrospect, I regret that attitude. Then, about ten years ago, I gave in and built a bird feeder. I put it out in front of the kitchen window so that we could begin our observations over breakfast. My wife Miriam and the children were delighted, and soon there was a telescope and a bird guide by the window.

The most exciting moment came when a surprise guest showed up: a middle spotted woodpecker. I particularly love this bird, because it is associated with ancient deciduous forests. Their habitat is threatened, because for these birds to feel at home, the beech forests need to have been around for a very long time. One reason for this is pretty straightforward: beeches that haven't reached the 200-year mark have smooth bark. It's only beyond this age that they develop wrinkles and grooves, as older people do – and only then that these woodpeckers can find purchase on their trunks. It turns out that this species of spotted woodpecker is also not very keen on constructing nesting holes. Perhaps, unlike other woodpeckers, they get a headache from hacking away at wood. Whatever the reason, this woodpecker either uses nesting holes vacated by other species or, if it absolutely must resort to its own handiwork (or beakwork, I should say), it works on trunks that are rotten, where the wood is already nice and soft.

And now this shy, rare bird was at my bird feeder. Until then, I had assumed that there were no middle spotted woodpeckers in my forest, so I was doubly glad: both for the bird and for the forest. The presence of this species is like an environmental seal of approval, and I'd been awarded one out of the blue. Of course, ever since then I wait eagerly for these special ambassadors of the woods to reappear. This they do regularly, because middle spotted woodpeckers are one of the few species of birds that remain tied to their territory even in winter.

Despite the joy of experiences like this, I want to look further into the question of whether feeding birds in winter is beneficial from an ecological perspective, because it certainly changes the rules in the bird world. When Gregor Rolshausen and his fellow researchers at the University of Freiburg looked at two different groups of blackcap warblers, they showed just how much. The birds, which are about the size of a tit, are easy to identify. They have grey plumage and sport a cap on their heads: black for males and brown for females. They spend their summers in Germany, and in autumn they fly to warmer places such as Spain. There they eat berries and fruit, including olives. But in the 1960s they established a second migration route that led not south but north, to the UK. The reason is that the British are great bird lovers, and they feed the birds in their country so well that they no longer want to fly south. The migration routes to this island nation are, of course, considerably shorter than to Spain.

Bird food and olives are so different that the original shape of the birds' beaks is not best suited for their new food source. Therefore, over the past few decades, the population of blackcaps that fly to the UK has begun to change, both visually and genetically. Their beaks have become

narrower and longer, while their wings have become
rounder and shorter. Both developments are adaptations
to life at the bird feeder. The redesigned beaks make it
easier to pick up seeds and fat; and while the wings are no
longer ideally suited for long-distance flights, they improve
the birds' manoeuvrability during short flights in the
garden. And because this group and the population that
continues to fly south rarely interbreed, a new species is
gradually being created – a species that is forming because
of winter feeding. You could call this a massive interference
in nature, but is it necessarily negative? Isn't the development of a new species something to be celebrated? After
all, increasing the overall number of species is always a win
for an ecosystem, and in this case it means the birds have
adapted to better fit a changing environment. It becomes
a problem, however, when the altered species then inter-
breeds with the original one, changing the genotype so
much that the original form of the blackcap, for example,
might cease to exist.[2]

This is what happened with many culturally manipulated plants, including fruit trees. There are hardly any
genetically pure wild apple or wild pear trees left: they
might well be on their way to dying out completely. The
reason lies in the cultural history of fruit, which stretches
back for thousands of years and has changed fruit
breeding. Because bees don't care which blossoms they
pollinate, they carry the pollen of fruit trees bred by
people to the blossoms of their wild counterparts. The
genetic material of the two gets mixed, and in the process
the offspring of the wild trees are changed. At some

point, pollinating insects will wipe out the last of the wild fruit trees, and all the trees will be hybrids.

Does it matter? We don't know, but we do know that something irretrievable is being lost. This happens in the animal world, as well. An auroch looks out at us from behind the eyes of every cow, albeit, genetically speaking, from a great distance. Although it's impossible to revive aurochs in their pure form, Heck cattle, which have been bred back to have at least a visual resemblance to aurochs, now graze in some nature reserves in Germany.

But, of course, there are other reasons to feed the birds, and here I come back to human emotions. It's not just the middle spotted woodpecker that has shown me how much enjoyment you can get from birds, but also Koko the crow. I introduced Koko in my book *The Inner Life of Animals*. We see the crow only in winter, and when we do, it's all about food. Our two horses, Zipy and Bridgi, are out in the field all year round, because fresh air helps keep them healthy. They're old ladies now, and they get a handful of grain every day so that they don't get too skinny. Koko used to peck undigested seeds out of their droppings, which I found less than appetising.

For some years now, my wife and I have been putting out a few kernels of grain on the rail where we tie up our horses so that Koko can get a somewhat more hygienic breakfast. I completely overlooked the fact that the crow was communicating with us. One day, it flew past me with an acorn in its beak and hid it in front of me in the grass. When it realised I was watching, it retrieved the acorn and flew on a bit further so it could bury the nut safely out of my sight. Only then did the crow fly back and collect its morning portion of grain. When I recounted the story at the breakfast table, my children enthusiastically suggested that I include it in my animal book.

You'd think after that I'd have been better prepared to notice the bird's behaviour, but unfortunately that wasn't the case. And so it was that when Koko left us a token of his affection, I was initially completely oblivious. I only realised what he was doing when Jane Billinghurst got in touch with me. Jane had already translated *The Hidden Life of Trees* into English, and now she was working on *The Inner Life of Animals*. To make my account more familiar for English-speaking readers, we were replacing some of the German references with similar ones published in English. To address the question of gratitude (and whether and how animals express this), Jane suggested I mention a BBC report from Seattle.

The story was about a girl called Gabi. When Gabi was four years old and eating outside, she would occasionally accidentally drop some of her food on the ground. In no time, crows snatched up the unexpected gifts. In due course, Gabi got into the habit of sharing the contents of her school lunch box with the crows, because she liked them. Eventually she started putting food out just for them. She set out containers of nuts, provided water, and scattered dog food in the grass.

This was the turning point in the relationship between Gabi and the crows, because now the birds began to bring Gabi presents too: little pieces of glass, bits of bone, and small beads or screws. They left their gifts in the containers after they had eaten the food. Over time, Gabi amassed quite an astounding collection.[3] I found the story moving and quickly agreed to include it in my book.

Then, the next time my wife and I trudged out to feed the horses (this was in December), we noticed a small apple on the rail. And now for the first time I realised what was happening. Koko had been leaving us gifts for years: we just hadn't noticed. We had often puzzled over the fruit, stones or sometimes bits of mouse that were laid out in the exact same spot where we had left the food, but it had never crossed our minds that these were gifts intended for us. In hindsight, we regret we hadn't noticed earlier what Koko was trying to communicate, and nowadays we're especially excited when the crow leaves us something.

So let's ask again: is feeding birds a bad thing? Aren't we interfering with natural processes? Without our help, Koko might have starved long ago, and another crow or another species of bird might have filled the vacancy that Koko left in the ecosystem. We've covered the pros and cons of the direct impact on the environment, but we haven't talked about another aspect here: empathy. Empathy is one of the strongest forces in conservation, and can achieve more than any number of rules and regulations. Think of the campaigns against whaling, or against the slaughter of seal pups – the public outcry was so loud only because we all empathised with the animals. And the closer the animals are to us, the greater our empathy.

And I mean that literally, which is one reason why in principle I'm not opposed to zoos, as long as the animals are looked after in an appropriate manner. People who get to meet animals up close feel more strongly connected to them, and are more prepared to do something to protect them. That's why I think it's a shame that private individuals (in Germany, at least) are not allowed to own wild animals. For species that are not threatened by extinction, this could, on balance, do more good than harm. Anyone who has had the kind of experiences I've just described will no longer curse the magpies in the front garden or support the shooting of ravens. Certainly an animal will occasionally be loved to death because it's not being looked after in the right way, but in the end the best way to protect nature is to ensure that people experience it.

One small tip, by the way. Birds don't just need food in winter: they're also in danger of dying of thirst. A little ice-free water in a dish can sometimes be even more helpful than bird food. We see this at our horses' drinking trough. The horses live outside in their pasture all year, even on freezing cold days. (They much prefer this, I hasten to add, to standing in a warm stable.) Anyway, water is a problem because the trough keeps freezing over. The only solution is buckets of warm water that we carry out either in a wheelbarrow or on our ATV.

And from time to time we see Koko and his companions helping themselves to a sip of cool, clean water from the trough after their meal of grain.

With other animals, however, feeding them in winter is not a good idea, because it can have a negative impact on the whole ecosystem. How that happens, and why trees today will never be free of wild boar, is too much to fit into this chapter. So let's begin a new one.

10

How Earthworms Control Wild Boar

It's often said that mild winters cause plagues of biting flies or catastrophic bark beetle infestations. We've already seen that explosions in bark beetle populations are more often caused by commercial forest practices, but I think it's worth taking a closer look at the impact winter has on the forest environment. Severe winters are characterised by weeks of hard frosts and at least some snow cover. Everything freezes. The top few inches of the soil are rock-hard, and life out there in the forest seems anything but a walk in the park.

Let's start by taking a look at how winter conditions affect smaller animals. Insects exploit the laws of nature to protect themselves against freezing. They use sugars they produce naturally to create a kind of antifreeze, and they empty their gut to minimise their water content, because tiny amounts of water don't freeze until temperatures fall far below 0 °C. Five microlitres of water, for instance, doesn't form ice crystals until the temperature reaches -18 °C. Despite this, the youngest bark beetles struggle to survive. If it stays cold for too long, larvae pass on to insect Nirvana without seeing spring. Their lives end not because they can't survive freezing conditions, but because water has entered their mouths and breathing tubes. Although fluids inside the larvae are protected against freezing temperatures, when the temperature falls water entering from outside their bodies freezes immediately. That's why the little ones survive particularly well when a thick layer of snow keeps out the worst of the cold. Because adult beetles don't have this problem – they can cope with up to -30 °C – bark beetles try to avoid breeding in autumn.

Yet mild winters are catastrophic for bark beetle larvae, too, because mild means damp. Think about it. What kind of weather would you prefer to be out in? A few degrees above freezing in the rain or a few degrees below freezing in the sunshine? I, for one, would prefer the latter. If temperatures are below freezing, you usually stay dry, which means it's easier to keep warm. Above 5 °C, moisture-loving fungi become active again. They get to work on the overwintering insects and start eating them while they dream.

Whereas overwintering bark beetles sit tight and wait for spring, most mammals are awake and active in winter, which means they need to eat all the time to maintain their body temperature. This puts them in the same predicament as the birds. So shouldn't we take pity on our four-legged fellow creatures too, and feed them? In fact, we're already doing that, at least with some species. Have you ever seen a fodder rack in the forest? Or perhaps a couple of wooden boxes filled with feed corn?

This food is supposed to help starving roe deer, red deer and wild boar survive the winter. But this supplemental feeding is not an act of altruism. The provisioning only benefits animals whose antlers or tusks would look good displayed as hunting trophies above the living-room sofa. No thought is given to foxes or squirrels. The good news is that it wouldn't be necessary anyway, as they are well adapted to their climes and have developed their own strategies to survive the cold months of winter.

Whereas squirrels cache food and sleep a lot in winter, deer have adopted a different strategy: they regulate their body temperature while spending the cold months of the year standing around dozing in the undergrowth. Researchers at the University of Vienna have discovered that deer can lower their subcutaneous temperature to 15 °C to save energy, which is amazing for large warm-blooded animals. According to the project leader Walter Arnold, this strategy is similar to hibernation.[1] Using this method of saving energy, deer can make the fat reserves they put on in the autumn last into the following spring. It's only the weak or sick animals that starve, which is a natural way of ensuring the genetic health of the species.

Winter feeding can in fact lead indirectly to starvation, particularly with red deer. This happened in Germany in the winter of 2012–13, when a lot of snow fell everywhere in the country. In the district of Ahrweiler, where I live, the deer population had increased so much that in the forest the animals were practically tripping over one another. Hungry deer were appearing in cattle shelters on farms and eating the cows' feed. A colleague of mine even sent me a photograph of a doe having a snack at a bird feeder. Of course, the hunting community began calling for a permit to feed them. Hunters even visited schools to solicit sympathy for the animals and gather support to influence politicians.

Many dead deer were found, which fired up the debate: were we really going to allow these noble animals to starve? When vets examined the animals, however, they made an unexpected discovery. They found the victims' stomachs were stuffed full of food, which ruled out hunger as the cause of death. But they also found enormous numbers of gut and stomach parasites, and it turns out it was these that had sealed the deer's fate.[2] Because of their increased numbers, the animals had more contact with one another and with contaminated faeces, which helped the parasites to spread widely. Once ingested, it was the parasites, not the deer themselves, that were the beneficiaries of the nutrients in the deer's guts, which caused many of the deer to get weak and die, even though they had been fed. Death by starvation, therefore, was an indirect consequence of the feeding programme.

The findings didn't change the hunters' views, however. They were still keen to help as many large mammals as possible survive, because that meant there would be something to shoot every night when they perched in their hunting hides. Overpopulation, however, also leads to stress, as deer fight over territory, and stress in wild animals leads to reduced body weight and, especially in roe deer, smaller antlers. So here's another unintended consequence, because what hunters really want is lots of wild animals with huge trophy racks. Ignoring the reality of the situation, hunters in Germany continue to try to fatten up weak individuals, which makes things even worse. And feeding deer isn't cheap.

The journal *Ökojagd* once estimated from hunters' reports how much they fed game, and came up with a ratio of 12.5 kilograms of corn per kilogram of bagged game.[3] That is way more food than the meat industry feeds animals in the mass production of meat. And because nature is the way it is, food is immediately converted into reproduction, so the animal numbers explode. The result is wild boar in vineyards, in front gardens, or even on the Alexanderplatz, the large public square in the middle of Berlin, because the forest is gradually becoming too cramped for the many game animals out there.

The hunters' interventions in the finely calibrated balance of nature create more losers: trees. Over millions of years, trees have developed a perfect strategy to keep large browsers away, but it no longer works when the animals are being fed by people.

The two most important native trees in Germany, the beech and the oak, produce very large seeds. A beech nut weighs only 0.02 grams, but that's quite a lot for the seed of a forest tree. Spruce seeds, which are an important food for squirrels, mice and many birds, are 4/100 the weight of a beech nut, but they're still very attractive to animals. Beech nuts, however, are what you might describe as real calorie bombs, not just because of their size, but because they contain almost 50 per cent fat. Acorns are even heavier, with an average weight of 4 grams. The fat content of an acorn may be only 3 per cent, but the carbohydrate content is 50 per cent, making acorns the top prize in the autumn food lottery.[4]

This lottery plays out on a three- to five-year cycle. In the off years, hunger is the order of the day for many animals – and that is exactly why beeches and oaks don't produce fruit every year. The breaks they take ensure that populations of wild boar, roe deer, red deer, birds and hordes of hungry insects don't come to depend on their bounty.

Wild boar in particular are really good at snuffling up these desirable seeds, and some years they eat every last one of them on the forest floor. Boar populations can then quickly triple, and a year later large groups of young pigs roam through the autumn leaves overturning every twig, stone and rotting stump. The following spring, no new beech trees sprout and no young oaks see the light of day, and if this situation persists for decades, forests begin to age.

When an ancient tree dies, grasses and bushes grow in the clearing that opens up around it, gradually creating a small grassy meadow where browsers now move in to eat all the small trees that try to sprout. Trees, however, know how to put a stop to that, and one way they put the brakes on meadow-making is to leave long intervals between the times they bloom in order to reduce the number of browsers in the vicinity. But that's not all. What use is it if only some of the trees take a break, while others are loaded with beech nuts and acorns? Wild boar go hungry only when there are no nutritious seeds anywhere for at least a few years.

And so the trees come up with a communal bloom strategy, with all trees of the same species getting on board. It's not enough, for example, for one stand of beeches to come to an agreement, which they do via their root and fungal connections in the ground. That form of communication works well – and, astonishingly enough, some of it is via electrical impulses – but the wood-wide web doesn't reach far enough for this particular purpose, because wild boar can roam long distances and simply find another stand of trees 10 to 20 kilometres away. Therefore, trees agree among themselves over long distances, and by long distances I mean hundreds of kilometres. We don't yet know exactly how this works, but the important thing is that, with the exception of a few rogue individuals, all the trees in a large area synchronise when they form fruit and when they take a break from reproduction.

In Germany these days, however, the deciduous trees' defence strategy is being completely obliterated by hunters. They feed wild boar not just in winter but often all year round, which cancels out the food shortage planned by beeches and oaks. When foresters in Baden-Württemberg examined the stomach contents of wild boar that had been shot, they discovered that on average through the year, 37 per cent of the food they eat came from hunters. In winter, this percentage increases to 41 per cent, which has had dramatic consequences.[5] In the cold months of all years except mast years the forest is mostly empty of food – as you would expect the stomachs of the wild boar to be. Without the hunters' intervention, therefore, many of them would starve, and then the population would once more be proportionate to the capacity of the habitat to support them. But boar no longer experience, and have to adjust to, periods of scarcity. At thousands of feeding stations they can help themselves any time they like, which just boosts their rate of reproduction even further. The German Ecological Hunting Association (ÖJV) calculated what that meant in stark terms for individual animals: in extreme cases in the Westerwald mountains, it came to as much as 780 kilograms of food per shot sow.[6]

The hunters try to conceal the true causes of higher populations. They blame farmers and their huge hog-heaven fields of corn. They argue that climate change, with its warmer winters, favours dramatic increases in population. They insist that they no longer feed wild game, because the practice is mostly no longer allowed, at least for wild boar. That's true only insofar as the word 'feed' has been replaced with the word 'bait'. But by bait we're talking about a small amount of feed corn put out in a clearing to lure animals within range of the hunting blinds, where they are shot. Thus, according to the official version, baiting actually leads to the reduction of wild populations, not their increase. But there is so much baiting going on that the population of wild boar is increasing faster than it can be shot; consequently, baiting is not having the stated effect, and the whole situation is getting out of hand. To say nothing of the fact that in most districts illegal feeding continues on a massive scale.

Out there in the forest, mostly out of public view, just about everything has been dumped that wild game might find tasty. At the beginning of my career, I found whole truckloads of tulips bulbs in a clearing. They were clearly not saleable, and therefore needed to be disposed of. Hunters probably thought, 'Why not make a virtue out of necessity?' and had the cargo transported to the forest. The wild boar seemed to have enjoyed it, because after a few weeks the bulbs had disappeared.

Apples that are too small or too light according to EU regulations, or that don't look the way they should, are also disposed of as food for wild game. An acquaintance told me that a hunter in her village once scattered tonnes of pralines that looked mouth-wateringly fresh. Hunters are basically doing what owners of large restaurants used to do decades ago, when it was normal to keep a barn full of pigs to process leftovers and recycle rejected chicken stew, duchess potatoes or pork and beans into fresh meat. It's the same when you feed wild game in the forest: the only difference is the pigsty, which is much larger and made of trees.

Meanwhile, the relationships that used to exist in ancient forests have been completely upset by foresters and hunters. Whereas there used to be very few roe deer per square kilometre, today there are, on average, fifty individuals. And as stags are animals of the plains, they were hardly ever encountered in the forest. Now, in many forests, there are an additional ten stags and ten wild boar per square kilometre, which means it's really crowded. The forests of central Europe have become a real zoo, which makes the heart of every hunter beat faster.

If there are hordes of browsers demolishing most of the next generation of trees, do our deciduous trees have any kind of a future at all? We mustn't give in to such gloomy thoughts. Fortunately, it's only a matter of time until conditions improve. For one thing, there are the wolves, which are slowly returning all over Europe to set things back on track like they did in Yellowstone. For another, the trees have other secret allies. Surprisingly, one of them is a creature that lives in the soil and can be very dangerous for wild boar: earthworms. *Earthworms*? Don't they stay peaceably in their tunnels, munching fallen leaves and excreting them as humus?

That's right, but they can still pose a danger to wild boar. At first, it's the other way round. Wild boar plough through soft ground with their plate-shaped snouts searching for food. One of the best sources for the nutrition they crave is indeed earthworms. Per square kilometre, up to 300 tonnes of them can live beneath the surface of the soil.[7] To put this into perspective, the weight of all the large mammals living in a similarly sized area (roe deer, red deer, wild boar) comes to only one-third of that. Incidentally, in the event of some sudden catastrophe, it would be much more effective for people to plough the land in search of earthworms instead of reaching for their guns.

Pigs eat earthworms, which in themselves are harmless, but when wild boar do this, they consume some stowaways at the same time. These are lungworm larvae, which develop inside the earthworms while waiting for a suitable host for the final phase of their life cycle. That host – and here I need to return to that state of emergency I just mentioned – could be a person. This means that if you do end up having to eat earthworms, you should cook them thoroughly first. If a wild boar eats a meal of lungworm larvae, however, the larvae travel through its bloodstream and end up in its lungs. There they settle in the boar's bronchial tubes, where they mature into adults that cause swelling and bleeding. The wild boar then excrete the eggs, earthworms ingest them, and the cycle is complete.

Thanks to their weakened respiratory organs, the bristly porkers become susceptible to a whole range of other illnesses, and there is a higher risk of mortality, especially among piglets. The higher the numbers of wild boar, therefore, the more earthworms carry lungworm larvae, which in turn means more infected pigs. The whole thing spirals upwards until, at some point, the boar population crashes. Fewer animals then means fewer excreted eggs and almost no infected worms. Lungworms, therefore, regulate wild boar populations, but there are still more little adversaries out there that the boar have to deal with.

A whole army of infectious agents has its eye on wild boar, including a large number of viruses. What exactly is a virus? Scientists don't include them among the living species of this earth, because they have no cells and can't reproduce or metabolise on their own. All they are is a hollow shell that contains a blueprint for multiplication. Basically, they're dead. At least, they're dead as long as they aren't docked onto an animal or a plant. Once they've made this connection, they smuggle their blueprint into the host organism, forcing it to create millions of copies of the virus. In the process of multiplication, mistakes are always made, because viruses, unlike cells, have no built-in mechanisms to fix aberrations.

All kinds of mistakes, however, mean all kinds of new variations on the virus. It doesn't matter that many of them go nowhere, because there's always something useful in a rubbish heap. That's how viruses adapt quickly to new conditions and can attack their hosts more effectively. New mutations, especially, have the potential to kill. Normally, it doesn't make sense for a virus to kill its host, because that would take away its opportunity to proliferate in the future. Only fresh mutations make mistakes like this, because they haven't yet adapted sufficiently to be able to exploit their hosts without killing them.

Naturally, the opposite is true for the host: a host that has a long relationship with a virus adapts over time so that the illness caused by the virus becomes relatively harmless instead of fatal. Chickenpox is an unfortunate example here. Europeans are well adapted to this illness, which usually strikes in childhood, but when the virus was brought to North America by white colonisers it devastated indigenous populations. In some areas chickenpox, in conjunction with measles and other diseases, killed up to 90 per cent of the population.

Animals face the same predicament. Globalisation creates similar conditions for them that humans found when they arrived on continents that were new to them. Illnesses that native fauna don't know how to combat arrive in a new territory packaged up with trade goods and or transmitted by imported animals and plants. African swine fever is one such disease. It's normally found in Africa, where bloodsucking soft ticks transfer the disease from one animal to another. In 2007 it was identified outside Africa in Russia, from where it spread to Europe, largely because people allowed this to happen. We don't exactly know how it was imported, but it probably arrived in a shipment of pork containing the virus. From there, the virus is likely to have spread through the illegal disposal of slaughterhouse waste and carcasses. The death rate for infected animals is extremely high: in fact, it's 100 per cent.[8]

Will it also affect wild boar? It's definitely a worry for individual animals or for a family of them. Wild boar are sociable and love to cuddle up together, so the infection can jump from one animal to the next. And even if all members of the family are not infected, the whole family suffers. Wild boar love their parents, children, brothers and sisters, and miss them when they die. However, swine fever is not necessarily catastrophic for of the forest ecosystem as a whole. It's difficult for the disease to spread naturally in central Europe, because there are no soft ticks here to act as intermediate hosts. Here, it's the unnaturally dense population of boar that makes direct transmission from one animal to another so much easier. So if the disease thins the population, there will be less contact between the boar, and the virus will not be able to travel further. Then the epidemic will end, there will be fewer boar trotting around in the forest, and beeches and oaks will have room to breathe again.

There are connections that we know to be true because they have been well researched (as in the case of wild boar and swine fever), and then there are connections that we assume to be true because they have been handed down for generations. But maybe the time has come to take a closer look at things many of us have always taken for granted.

11

Fairy Tales, Myths and Species Diversity

ALTHOUGH WE'VE LOOKED at a range of natural connections – some of which are quite surprising – I haven't mentioned others that might seem much more obvious to you. And there's a good reason for this: these connections exist only in the imaginations of our ancestors.

Since ancient times, the fruit set of beeches and oaks has been used to forecast the weather. There are old German country sayings like 'Many beech nuts and acorns and winter will be harsh', or 'Many acorns in September, much snow in December'. To try and get to the reality that inspired such sayings, we have to ask why a tree would do this. How could forming lots of seeds help a tree survive a harsh winter? And what would the indirect consequences be for that tree?

I'm afraid I don't have answers. All we know is that both oak and beech trees agree on a time to bloom together, so that every few years they produce an enormous amount of seed. As I mentioned earlier, they do this to regulate the size of the plant-eating population, by ensuring that browsers can't rely on a constant supply of food every year. But the strategy has nothing to do with winter.

There's another point to be made here. Flower buds (just like leaf buds) are set the previous summer. So if a tree matched its seed production to the winter temperature, it would have to know more than a year in advance what this was going to be and plan accordingly. In fact, beeches and oaks have no better ways to predict winter weather than we do. What trees can do, however, is register shorter days and falling temperatures. They use this information to decide when to drop their leaves before the first heavy snowfall. But frequently they don't even get this short-term forecast right, as you can see when winter arrives early. When snow falls in October, as it often does, branches with green leaves still on them break under the heavy weight of fresh wet snow, which is a painful lesson for the trees. At least if this happens to them when they are young, they can learn from their experience and drop their leaves a little sooner in the future. But they only drop their leaves early as a precaution: their decision doesn't have anything to do with improved forecasting. There simply isn't any evidence that trees can forecast the weather a year in advance.

So much for the plant kingdom, but what about animals? There's also folk wisdom that says that squirrels can predict a harsh winter: if they're particularly busy gathering food and laying up large stores of acorns and beech nuts, then that means the winter is going to be particularly harsh. Really? I think you probably know the answer yourself. Of course these pretty little creatures don't have a sixth sense for what the weather will be like, any more than trees do. Their drive to collect nuts is simply a question of supply. When trees produce a lot of nuts, the little rascals hide a lot of them. In the years when the trees have agreed to take a break and there are barely any nuts on them, the animals find far fewer of them, and we're unlikely to see them squirrelling them away.

Then there are those folk sayings that straddle myth and reality. They express relationships that exist, but the explanations they give are incorrect. The classic one for me is the conventional wisdom that ticks like to live in broom. This shrub is widespread everywhere in Europe where the Atlantic ensures cool summers and mild winters, as it does in the Eifel, where I live. Broom is so common here that it defines parts of the landscape. In spring, the shrubs are totally covered in golden yellow flowers that resemble butterflies' wings. The blooms are so thick, there's not a spot of green to be seen. Large broom hedges infuse the landscape with a golden glow, and that's why the plant is called Eifel Gold.

So, do ticks really like broom? In fact, all parts of the plant are poisonous, and not only to people. Substances in the branches, flowers and leaves discourage browsing, and both deer and cattle generally give the hedges a wide berth. Where there's a dense population of wild game, most of the other, tasty shrubs get eaten. Broom, therefore, has an advantage over the competition and can spread unchecked – and that's just what it does, stubbornly and successfully. This particular shrub has developed a number of different strategies for spreading its seed. In the heat of the midday sun, for example, the seedpods burst open with a loud crack, scattering seeds in all directions. The round seeds roll downhill easily to travel even further.

But the plant doesn't stop there. Broom also employs ants, those secret sovereigns of the landscape. Ants distribute the seeds and help Eifel Gold get established in every corner of the countryside, even in forests. It's usually too dark there for the broom seeds, but time is on their side. They can lie dormant in the humus for more than fifty years, until one day a storm topples or a forester fells some trees. Rays of sunlight hit the forest floor and gently wake the slumbering seeds. They quickly sprout and grow into bushes up to half a metre tall in their first year. Only young trees or other bushes like raspberries cramp their style, but munching roe deer come to their rescue. They quickly eat this fresh green growth, removing the annoying shadows falling on the young broom bushes.

While the deer are munching away, a special kind of hitchhiker tumbles from their fur: ticks. The ticks that drop off are particularly large ones in the final stage of their life cycle. They have been sucking up blood one last time before loosening their grip and falling off so they can crawl into the nearest bush to lay their eggs and die. The young ticks hatch, are scraped off their bush by passing mice, and pick up where their mothers left off, continuing the family business of sucking blood. Then they, too, let go and fall off after their meal, grow and moult. Finally they sit tight, hungry once more, in the surrounding vegetation – the broom, for example – and wait there for larger mammals (and even people). And that's why there are always large numbers of ticks where there are large numbers of deer, and it is deer that ensure broom can spread freely. What ticks love is not broom but warm-blooded hosts. Broom is simply a species that also profits from the presence of browsers, and broom and ticks appear together in habitats where there are overly high numbers of deer. The ticks and the broom depend on the deer, but not on each other.

Trees can achieve great things together without meaning to, even when their achievements have nothing to do with survival. Every year in autumn, a drama plays out that makes me think of the merry-go-round in a children's playground. Someone gives it a spin, while the children sitting on it stretch their legs out in front of them away from the centre. When they draw their legs in, the merry-go-round spins noticeably faster; when they stretch them out, it slows down again. It's doubtful whether trees have fun on merry-go-rounds, but every year they do something similar. When deciduous trees in the northern hemisphere drop their leaves all at the same time, we all spin a little faster and the days get shorter. You don't believe me?

We're talking only tiny fractions of a second, which, I have to say, are barely perceptible because of the influence of other global processes, but the change is measurable. Most of the land mass of the earth lies in the northern hemisphere. It's therefore here where we find most of the trees. When deciduous trees lose their leaves, their discarded foliage ends up about 30 metres closer to the centre of the earth (that's the average difference between the treetops and the ground). And the effect of this shift in weight towards the centre is similar to the effect of children drawing in their legs on a merry-go-round. Then in spring, when new leaves grow, the exact opposite happens. The fresh, water-filled leaves shift a lot of weight up higher or, to put it another way, further away from the centre of the earth, slightly slowing us down again. A more entertaining way to look at it is that, thanks to the trees, we get to ride on a merry-go-round. However, because this only makes a difference of fractions of a second, and because there are other processes that also affect the earth's centre of gravity, such as tidal currents, we may as well put this spinning phenomenon in the basket of half-truths that blur the line between fact and fancy.

There's a very different kind of myth surrounding biodiversity. When we save individual animals or plants, we really believe we're doing something good for the environment. Yet this is rarely what actually happens, usually because when we have to change conditions in the environment to ensure the survival of one species, we put the survival of many others in jeopardy. But I'm getting ahead of myself.

When we see just how multi-layered the interactions between different species are, we have to ask once again whether we'll ever be able to fully comprehend the connections in nature. The examples we've discussed so far involve just a few animals influencing each other in highly complex ways. Imagine a juggler starting with two balls in the air, but one by one, more and more balls are added. In nature, every time a new species enters the arena, things become much more complicated and difficult to follow. According to current estimates, there are 71,500 of these 'balls' in Germany – including animals, plants and fungi – and to date 1.8 million species have been identified globally.[1]

But it's even more complicated than that, because there are many animals and plants we haven't discovered yet. In 2014–15 a report estimated that in the Amazon a new species was being discovered every 1.9 days. Recently, I was talking to a researcher who, incidentally, is herself a member of an endangered species: entomologists. There's simply not enough money available to research beetles, flies and other insects, she told me, and, more importantly, not enough young scientists entering the field, which means that even in Germany there are still a good many blank spots on the species map. To the 71,500 known species in Germany, therefore, you can add an unknown number of species whose effect on the ecosystem is, inevitably, unknown. It's clear we can never apprehend everything about nature correctly – but then in my opinion we don't need to.

In the examples we looked at in previous chapters, it's easy to see how fragile the system is, and what the consequences are when just a single species falls off the map. Knowing this, we must strive as hard as we can to preserve intact landscapes or leave them to their own devices. But what does intact mean? Who should we trust here?

In Germany, forest agencies and owners claim that commercial forests are good for biodiversity. The third National Forest Inventory, published in 2014, takes great pride in the fact that the average tree is now seventy-seven years old. Bravo! The brochure, published by the Federal Ministry of Food and Agriculture, also extols the ecological importance of old trees and implies that on this score everything is hunky-dory.[2] The tree sap hoverfly, *Brachyopa silviae*, would certainly dispute this if it could. This tiny fly was first discovered in 2005. It has only been spotted six times in the world, so you could say it is extremely rare. And there is a reason for that.

Even though it has wings, it seems that this fly doesn't travel very far, but prefers to remain in undisturbed forests, where it feels at home. Here, the fly looks for wounds under tree bark that are weeping sap, its favourite food. Or rather, this weeping sap provides the substrate for its favourite food, because the sap is nutrition for bacteria and other micro-organisms that form a slimy mat on top of it, and it's here that the tree sap hoverfly likes to graze. But weeping patches like these can be found only in trees that have at least 120 years under their belt (perhaps I should say, their bark). *At least* 120. If the official brochure is happy with an average

age of seventy-seven, then we should worry about the tree sap hoverfly.

Dr Frank Dziock discovered this fly only by chance.[3] He had set out insect traps in flooded areas to catch hoverflies, because he wanted to find out how they reacted to high water levels. At first, Dziock didn't realise that he had found something remarkable in his trap. Then he noticed two spots on the back of one of the flies. No other known fly sports these spots, and so he knew what he had here was an as-yet-undiscovered species.

This fly needs wounded old trees. Old trees, however, are highly endangered in commercial forests, where damaged trees are targeted for thinning. The long-term goal is to let only perfect specimens of beech and oak grow larger and older so that their valuable wood can be harvested. Too bad for the flies: their needs are completely ignored. It's true that a few trees are left scattered about for environmental reasons, but if all the others around them are chopped down, these survivors won't grow to be very old. Without the typical damp, cool microclimate of the forest they'll struggle, and direct sunlight will be heating up the ground all around them. To make matters worse, their root and fungal networks, which help support old and sick trees, will have been destroyed. This network is crucial for forest health, so let's take a closer look.

In my book *The Hidden Life of Trees*, I talked about the forest internet, or the wood-wide web, as the journal

Nature felicitously described it. This network is made of fungi which grow their filaments through the soil and connect trees and other plants with one another. Fungi are remarkable fellows. They don't belong in either the plant or the animal kingdom, but they have much in common with animals. They don't do photosynthesis, so they have to get their food from other living entities. Like insects, they have chitin in their cell walls. A few of them – slime moulds, for example – can even travel from one place to another. Not all of them are friendly, however. Honey fungus, for example, attacks trees to get to the sugar supplies and other delicacies stored inside. Often it kills its victim and then moves along the ground to the tree's next family member. But trees are not defenceless in the face of attack by fungi or insects: they heed warnings from each other, including scent signals that contain information about which villain the trees are up against. The appropriate defensive compound can then be stored under the bark to spoil the appetite of hungry insects or mammals.

Unfortunately, the wind often blows airborne warnings only in one direction. This is where the roots come in. They connect with the roots of other related trees and transmit important news bulletins via both chemical and electrical signals. But this root network can't reach every corner of the forest, and sometimes the connection is broken when an ancient tree dies. Fungi help bridge such gaps. Like the fibre-optic cables of our internet, their subterranean filaments carry messages from tree to tree so that the whole forest soon knows what to expect. However, this service isn't free: the fungi tap Beeches, Oaks & Co. for up to one-third of the sugar and other carbohydrates they produce by photosynthesising. That is a sizeable chunk of energy, and about the same amount as the tree uses to grow wood. (The remaining third is converted into bark, leaves and fruit.)

Anything that makes such high demands must also deliver dependably. And fungi seem to have mastered this, even though it's a tricky task. The wood-wide web often suffers massive disruptions. For example, in winter wild boar roam the forest ploughing deep furrows in the ground as they search for beech nuts, acorns or the nests of mice. Inevitably, this activity disrupts fungal connections for many square metres. It's no problem for the fungi, though. As a fall-back option, they extend many filaments parallel to one another, and simply switch the connection to adjacent threads. Incidentally, that's why when you're out collecting ceps, boletes or chanterelles in the autumn it doesn't matter whether you twist the mushrooms or cut them off (a perennial bone of contention among nature lovers). Underground, any damage is quickly bypassed.

In addition to sharing information and transporting sugar from one tree to another to support weak members of the tree community, the fungi also help trees reach essential minerals. For example, when trees suck up phosphorus compounds, they soon exhaust the supply within a few millimetres of their roots. So it's a good thing that fungi connected to the larger network spin their filaments around the trees' delicate feeder roots to extend their reach. That way, all the nutrients the trees need can be delivered to their door even from distant parts of the forest floor.

Fungi can live to a ripe old age but, like every living thing, they start very small: as spores. And spores have a big problem. If they drop directly down from the cap of the fruiting body (the mushroom), they fall onto ground already occupied by their mother, which means they can't colonise new territory. Billions of microscopic specks fall out of a single cap with travel on their minds, which is a real problem on a forest floor where normally no breezes blow. And this is where the special construction of the fungal fruiting bodies comes into play.

The fruiting bodies are mostly constructed as a stalk with a cap on top, and there's a good reason for that, as the biomathematician Marcus Roper at the University of California, Los Angeles discovered. The spores trickle out of openings on the underside of the cap, where they are sheltered from the rain so they don't get wet and clump together. The cap itself transpires water vapour, cooling the air around the mushroom just a little. The cooled air around the edge of the cap sinks slightly, taking the spores with it, and then warms up again in the surrounding air. Both the warmed air and the spores now waft out and up about 10 centimetres above the cap.[4] All it takes is a tiny breeze to carry the little passengers away, and the survival of Ceps & Co. is assured.

With any luck, one of the tiny spores lands on ground that's not yet spoken for. There, it extends a few of its small threadlike filaments (hyphae) and waits for signals from plant roots. If there's no chemical call in the vicinity, the spore retracts its hyphae. It has enough food for multiple attempts at connection.[5] If it succeeds in making contact with the plant it's looking for – a beech tree, for example – it can be the beginning of a very long life. Fungi can live as long as trees. Ancient honey fungus networks have been found underground in North America. The record holder is a fungus belonging to the species *Armillaria ostoyae*. It is 2,400 years old, and has spread to cover almost 9 square kilometres.[6]

When it comes to researching the world of fungi we've only begun to scratch the surface, and there are untold numbers of secrets hiding under every step you take outside. But there are also industrious creatures in the trees themselves that require very specific conditions to survive. No, I'm not talking about bark beetles, which are simple souls when it comes to their dietary needs. All they require of trees is that they must show signs of weakness so that they can't defend themselves. If that's the case, all the beetles need to do is start munching on their bark and cambium – the layer of living cells between the bark and the wood. And because these conditions are almost always met within the range of any given tree (and the beetle that specialises in eating it), there are hardly any endangered species of bark beetle. Things are quite different, however, with specialists. They're so demanding you could almost call them picky. And in the case of *Tenebrio opacus*, a kind of mealworm, even 'picky' might be an understatement. This insect feels comfortable only after a whole list of requirements have been ticked off.

Imagine there was once an ancient beech forest where a pair of black woodpeckers had taken up residence. They claimed many square kilometres of forest as theirs and fashioned an extended series of living quarters. Because the wood was hard and difficult to work, the birds took their time. Unlike other woodpeckers, black woodpeckers prefer to set up house in healthy trees, and who can blame them – would you want to live in a house that was crumbling around your ears? Healthy beeches, however, are very hard, even for woodpeckers. Unlike a human brain, a woodpecker's brain sits firmly in its skull, so that it doesn't bounce back and forth when the bird is using its beak to deal staccato blows to a tree. And to prevent concussion, there's a special springy support behind its beak that cushions the blows before they travel to its skull. Despite this, fresh wood is simply too dense. But the black woodpeckers are patient. They start their construction project by hacking out the entrance in the tree's outer growth rings. They then abandon the site, sometimes for years.

In the woodpeckers' absence, fungi take over. They're on the job mere minutes after the first turn of the shovel – or in this case, the first blow of the beak. There are multitudes of their spores in every cubic metre of air, and they immediately land on the site of the damage. Fungal growth quickly appears and starts to decompose the wood by eating it alive. The wood becomes soft and mushy, so that, after years of waiting, our woodpecker couple can finally return to building their home without getting a headache. And once the hole is finished, the woodpeckers can start a family. Things don't always work out, however. Other birds are quick off the mark and try to move in uninvited. Black woodpeckers can get rid of shy stock doves with a few energetic threats. Jackdaws, on the other hand, stay put and keep possession of the hole, forcing the woodpeckers to start all over again. Luckily, they usually have a number of living quarters to choose from, partly because males and females prefer to sleep in separate bedrooms.

Over the decades, these holes in the trees continue to rot, and the floor in them slowly sinks. When the young can no longer reach the opening out of which they must one day fly, these holes have now become too deep for black woodpeckers. So they move out, and now the retiring stock doves get their turn after all. They simply raise the level of the floor by adding nesting material − a solution that doesn't occur to the woodpeckers. As the cavity and the entrance hole continue to decompose, the entrance eventually becomes wide enough to admit owls. They like to use the holes, which by now have grown to an impressive size, and often take up residence for many years. One or two little yellow-necked mice also make themselves comfortable in the dry warmth inside the tree, dropping morsels of food and shedding flakes of skin.

And this is where our picky mealworm enters the picture. It's not until now – after the succession of hole-squatters has moved in and out – that it makes itself at home. The reason is its distinctive food preferences. Mealworm beetles, or, to be more precise, their larvae, love the mixture of fungi-processed crumbly, mealy morsels of wood, insect remains, feather fluff and flakes of skin, along with all the odds and ends the squatters have allowed to trickle down from above. Bon appétit![7]

It should come as no surprise that populations of this mealworm and similar species are now endangered. In commercial forests trees that rot for decades in the way I've just described are not particularly prized. They're often cut down and sold the moment initial woodpecker damage is noticed, and before internal rot makes the wood less valuable. True, here and there individual trees are left standing to do at least a little bit for species conservation, but such lonely outliers are not much good on their own. You need a lot of these kinds of holes to safeguard populations of all the living things that are part of this delicately balanced community.

So the mealworm beetle faces the same fate as the hoverfly, and there's only one way to support species like this and others. Instead of attempting a rescue mission by saving scattered individual trees from being harvested, large areas of forest should be taken out of commercial forest production completely. The claim that well-regulated forestry can do a good job of combining commerce and conservation across the whole forest should be banished to the realm of myth and legend immediately.

Just as trees are not defenceless when bark beetles attack, so they don't have to stand idly by accepting whatever the climate throws at them. And it's not only that they are capable of enduring an enormous range of temperatures: they can also actively influence the weather.

12

What's Climate Got to Do with It?

TREES ARE NOT ENTIRELY at the mercy of variations in climate, at least not if they live together in large forests and operate as a community. There are limits to what they can do, of course, but if trees work together, they can not only regulate the humidity and air temperature in the forest but also exert influence in other ways for miles around. A recent report from an international group of researchers looking at how commercial practices in Europe have changed forests made me reflect on this.[1]

The focus of their study was the switch from the deciduous forests of old to modern coniferous plantations. What the scientists working with Kim Naudts at the Max Planck Institute for Meteorology were most interested in was how trees reflect light. Deciduous trees are lighter in colour than conifers. On a hot summer day, moreover, ancient beech forests, which once dominated central Europe, transpire up to 2,000 cubic metres of water per square kilometre, which cools the forest air for a long way down below their crowns. The dark green crowns of conifers, on the other hand, absorb more solar radiation, which has a warming effect.[2] Conifers are also thriftier with moisture, so the air in coniferous forests is drier, and the way the trees manage water intensifies the warming effect of their dark needles.

The focus of this chapter, however, is not the effect of forestry on climate change, but whether there's a reason conifers behave as they do. Whether or not they're being grown commercially, these are not trees bred specifically for plantation life: they behave the same as wild trees growing in ancient forests in cooler climes, which is where they originated.

And this is precisely the part of the world where their behaviour could be advantageous. Out on the taiga, summers are short, often lasting only a few weeks. That means the conifers there have little time to grow, let alone form cones and propagate. It might well be that by increasing the ambient temperature, these forest ecosystems are simply trying to extend the warm season by a few days, buying themselves time that could be crucial to their success. This sounds logical, but right now it remains simply speculation.

Their strategy for surviving winter is further proof of how desperately spruce and pines need every warm day they can get. Unlike deciduous trees, these conifers keep their narrow, pointed leaves on their branches all the time, so that they can get to work photosynthesising immediately conditions are favourable. In central Europe, this can be as early as the end of February or the beginning of March, when the deciduous beeches and oaks are still deep in their winter slumbers. As soon as the sun warms the air — and the dark crowns of the conifers — spruce and pines begin producing sugar.

That sounds logical, too, and can be seen every year on sunny days as winter wanes. Yet it's only half the story, for another feature of conifers appears to contradict the processes I've just described. Throughout the endless forests of the taiga, there are other substances floating in the air: terpenes emitted by spruce and pines. When fresh, tangy scents greet your nostrils as you're out walking in a coniferous forest, these are the substances you're smelling.

The hotter the sun, the stronger the smell, and this connection is probably not coincidental. Researchers have discovered that water droplets attach themselves to the scent molecules the trees emit. Clouds don't just happen. In the air water molecules often bump into each other, but instead of sticking together, they part. If that happened all the time, it would hardly ever rain. For precipitation to happen, a large group of water molecules has to clump together and get heavy enough to fall as raindrops.

These clumps don't form unless there are small particles wafting through the air that the water molecules can adhere to. There are lots of these particles out there in nature: ash from volcanoes, dust from the desert, tiny salt crystals from the ocean, but above all, particles actively emitted by plants. And here our conifers play an important role. They discharge enormous quantities of terpenes into the air. The hotter it is, the more they emit. The terpenes would probably be just a fresh, tangy smell if it weren't for a second component: cosmic rays, which are tiny particles from the universe. These rain down on us constantly, even passing right through our bodies – in fact, they're passing through you right now even as you're reading this book. These rays make the terpenes ten to a hundred times more effective than they are naturally, because they make the trees' discharges clump together. When terpenes clump, water attaches to them very easily.[3] And so the endless coniferous forests of Siberia and Canada summon, or I should say create, rain all by themselves.

When a forest creates clouds that don't drop rain, that's still quite an achievement. The high swirling mists cool the air considerably and slow the rate at which water evaporates from the ground. But if the trees manage to conjure into being not just a couple of clouds but a substantial thundercloud, that's like hitting the jackpot. Even a small thundercloud can easily hold 500 million litres of water.[4]

Now, of course, we have a problem. On the one hand, coniferous forests heat up the air with their dark crowns, which makes them ready to start growing more quickly in spring. On the other, they cool down the air by forming clouds. Is this all just an accident? One of nature's whims? Am I perhaps seeing connections where there aren't any?

Maybe taking a look at the seasons in which these phenomena occur will help. When the first warm days of spring allow spruce and pine to get a jump-start on the growing season, it's still relatively cool. Thanks to the dark needles, the sun can make the air a tiny bit warmer, immediately heating the trees' tissues and helping them kick into gear much earlier than deciduous trees, which must first go through the laborious process of growing new leaves. 'A tiny bit warmer' really isn't much here: all it takes is temperatures above -4 °C. Then spruce can start producing sugar, but they're not emitting many terpenes.

It would be counterproductive to put up an enormous sun screen made of mist this early in the growing season. At temperatures up to 5 °C, the spruce is metabolising but not increasing its girth, which means the tree is basically marking time. Production doesn't kick into high gear until temperatures exceed 10 °C. At this temperature, sunshine is being converted into sugar, new wood is being grown, and energy is being invested in extending the length of its branches and roots. It therefore doesn't make sense to do any cooling until it gets really hot later on in the summer. Conifers begin to suffer significant damage once temperatures exceed 40 °C.[5]

Does this sound too warm for Siberia? Here it gets so cold because it's so far away from the moderating influence of the ocean. In winter, water in the ocean acts like a heater, and in summer, it performs the role of air-conditioning, as the air passing over the water is either warmed up or cooled down. In the interior of the continent, this effect is barely noticeable, which is why temperatures far inland are so extreme in both winter and summer. Therefore, it's only logical that the conifers which are so widespread in these regions have developed systems to warm up as well as cool down – and the latter also ensures that, every once in a while, it rains.

When you look at photographs of the taiga, or perhaps even travel there yourself, you'll notice that spruce and pines are by no means the only trees in the landscape. The deciduous faction is well represented, too, especially by birches. If spruce manage the somewhat adverse climatic conditions remarkably well, then birches must suffer correspondingly badly. Deciduous trees emit far fewer organic substances, and at the onset of spring there's no dark foliage to warm their cold and clammy little trunks. They get a far later start in spring than conifers. In addition, their leaves need to be completely replaced every year, which costs them extra energy.

So what's the advantage to being deciduous? In fact there are two. The first is that they are less affected by drought. In winter, deciduous trees lose less water than conifers, as on those few days when it warms up, they don't transpire, because they don't have any green on them. The second concerns their offspring. The seeds of deciduous trees like birches, poplars and willows travel much further than seeds from cones, and following a forest fire can arrive quickly and be the first to form new forests. But the older the forests become, the more spruce and pines prevail. Then the forest gets darker, and the light-loving deciduous trees disappear once more.

*

Every tree has its ecological niche and preferred climate, and central Europe has a few peculiarities that make the lives of these gigantic plants quite a challenge despite the relatively mild temperatures. The climate here is described by the cryptic sequence of letters Cfb. These letters stand for a temperate climate with warm summers and an even distribution of rainfall throughout the year.[6] That sounds good: moderate, warm and moist. But more important than these three adjectives are the extremes: heatwaves above 35 °C and cold snaps below -15 °C are a challenge for native trees.

Below about -5 °C, trees contract, which is to say they get skinnier. That's not because the wood itself shrinks, as a purely mechanical process wouldn't reduce the diameter of the tree by much, and shrinking can see the tree lose up to a centimetre. What's happening, it turns out, is that water is being drawn further inside the tree, a process that is reversed on warmer days.[7] Clearly, trees do not shut down completely when they take their winter break.

Even the oak, that champion of extremes, can be pushed to its limits in severe cold. An oak can survive these conditions only if it has grown old without any wounds to its trunk. If it is unscathed, its wood is flawless and evenly structured. Woe to a tree that has had hungry deer taking bites out of its bark, or tractor tyres disturbing the base of its trunk. If the oak has been damaged, it will have had to seal off its wounds and cover them with new bark. And that's when the problems begin.

Normally, a tree's woody fibres are organised in a uniform vertical pattern to avoid stress in its trunk. When a storm bends the tree over a bit, this vertical arrangement ensures it is flexible enough to sway back and forth. Wounded trees, however, have other priorities, at least around the wound site. They have to grow new bark over the exposed wood, which causes them to call upon the cambium. This crystal-clear layer divides to form new bark cells on the outside and new woody cells on the inside, which is how a tree increases its girth every year so that it can support its growing crown. In its haste to heal, the tree can forget its regular growth pattern in the area around the wound, and thick swellings form under the new bark.

The wood thickens because the tree is rushing to heal itself. If it dallies, fungi and insects will have a better chance of brazenly pushing their way inside. In the chaos, the tree has no time to worry about neatly organised fibres, and at first that doesn't matter. After a few years – trees are really slow – the task is complete. The wound has healed over, though there will always be a thick scar to show where a deer or tractor hurt the tree. But just because the trauma is over doesn't mean it's forgotten. Now it all depends on what other stresses come along. One day, cold temperatures arrive, and our veteran tree is at a distinct disadvantage, as the damp sapwood inside it freezes solid, and the ice threatens to shatter its trunk.

Healing the old wound has resulted in a chaotic bundle of fibres that exert varying amounts of pressure on the wood around them when they freeze. On clear, frosty nights, cracking noises reverberate like rifle shots through the forest. These are not hunters at work but the oaks themselves. Their woody tissue fails around the old wound and splits open so abruptly that the sound travels for miles, in a phenomenon known as frost cracking.

In hot summers, other problems arise. Normally trees regulate their microclimate themselves. They all sweat together, as we saw when we looked at their enormous use of water on hot days. The moist air brings the temperature down by a number of degrees, and the trees maintain the temperature they enjoy. However, if it's dry for months on end, at some point reserves in the ground are used up. The first thirsty trees send out a warning over the wood-wide web and advise all the others they'd better be frugal with the last few drops.

If it remains dry and the sun burns hot in the sky, the only thing to do is jettison leaves. At first, just some of the leaves turn yellowish brown and fall to the ground. This rids the trees of some of the surface area that transpires, but it also means sugar production is drastically reduced. Hunger, the lesser of two evil, replaces thirst.

If the rains return in mid- or late summer, it will be too late to grow new leaves: this needs to happen by the end of June. With reduced photosynthesis, the trees must tap into reserves they have set aside to grow new leaves next year, and these reserves will get depleted before spring rolls around. If the trees now get attacked by pests, they will have barely enough energy to fend them off. All this is made much more dangerous when heavy machinery used by modern forestry compacts the soil around the trees, which then can't store much water because the pore spaces in the soil have been flattened under tonnes of weight. For the trees that remain when the forest's water tank is crushed as other trees are harvested, thirst becomes even more of a problem in hot summers, and the situation is exacerbated by the greenhouse effect.

Current climate change is heating up tempers as well as the atmosphere. For some, climate change means the end of the human race and all life on the planet; for others, it's a natural phenomenon and the climate has always changed. The latter point of view is clearly true, but not very helpful. We all know that ice ages and warm interludes have come and gone over vast periods of time. However, even though I believe that human-made climate change is real and already having dramatic effects, I'd like to start by looking at arguments on the opposing side. Let's take a look at the natural cycles of carbon dioxide on an appropriately vast timescale.

In the Cambrian Period, about 500 million years ago, there were already vertebrates very distantly related to us. They had to deal with levels of carbon dioxide that sound like something out of science fiction. Whereas we have pushed the amount from 280 ppm (parts per million) to more than 400 ppm, the quantity in the Cambrian was more than 4,000 ppm. Then it sank, before rising again massively to around 2,000 ppm about 250 million years ago. Why didn't the earth collapse from heatstroke?

Many scientists predict that even if we add a few hundred ppm to pre-industrial carbon dioxide levels, life should be basically impossible. But clearly it was possible, or humans would never have come into existence. It's a question of the speed at which change happens, and thus the chances species have to adapt, that decides whether climate changes like these are catastrophic or benign.

Most of the time, the rate of change in our planet's history has been slow. Among other things, it's tied to plate tectonics and continental drift. When continents are moving quickly, and the African plate, for example, is being rammed under the Eurasian plate, mountains rise where the plates collide. The higher the mountains tower over the land, the faster the rock of which they are made erodes. You can see this in the Alps, where piles of small loose stones cover the feet and lower slopes of the mountains. This scree is weathered down to sand and dust that is then washed away by water and deposited elsewhere, along with the carbon dioxide captured in the redistributed material. In times of low tectonic activity there is a correspondingly lower supply of newly eroded rock. This is when volcanoes enter the picture. They spew out molten rock, and the intense heat releases the carbon dioxide bound up in it. In tectonically quiet times, more carbon dioxide is released by volcanic activity than is recaptured in eroded rock. When the earth is pushing the continents together with great force, the situation is reversed.

Does that sound complicated? Yes, it does, and yet it's important to understand these vast cycles to get the big picture. If volcanic activity were not releasing carbon dioxide out of rock and returning it to the atmosphere, we'd be facing a completely different problem. At some point, we'd run out of carbon dioxide, and that would be fatal. The reason why oxygen is our most important elixir of life is that the cells in our body need it to burn carbon compounds. Without carbon, the purest breath amounts to nothing. Plants fish carbon out of the air around them and store it in the form of sugar and carbohydrates. It is vitally important for us that we don't run out of carbon dioxide.

But that's exactly what the far-distant future seems to hold. For hundreds of millions of years, leaving aside the fluctuations, the concentration of carbon dioxide in the atmosphere has been steadily falling. And the warmer the world gets, the more this process speeds up, because warmth increases the rate of erosion, and therefore the degree at which the gas binds to tiny particles.

'Hundreds of millions of years' are the important words here. Yes, over the very long term, the concentration of carbon dioxide can and probably will sink further, but it won't completely disappear, because volcanoes will always be releasing it. And life will adapt to lower levels, as it has always done. Much more important are the relatively rapid short-term changes that upset the finely tuned balance, such as the current increase in carbon dioxide concentrations resulting from the burning of fossil fuels: we are taking carbon dioxide out of the ground and releasing it into the atmosphere at an unnatural rate. Such rapid short-term changes have happened time and again in earth's history, and every time, many forms of life died out abruptly. At the moment, we're staring at rising carbon dioxide levels like a deer caught in the headlights, while the thing that should concern us most is the rate of change. Higher temperatures are not in and of themselves bad, as long as nature has time to adapt.

The problem is particularly obvious with trees. Populations of trees move very slowly. They can't simply shift a few hundred kilometres north every few years, even if the wind or birds carry their seeds for them. When a jay transports a beech nut in that direction, the seed must sprout, grow and then at some point, when it is a mature tree, produce offspring off its own. There are therefore long pauses on these journeys north that last for centuries. And so the average rate of advance is about 400 metres a year. This means that escaping north takes thousands of years – time that right now Beeches, Oaks & Co. don't have. And the species that are already settled in the North need to figure out how they are going to manage as conditions change.

In these times of climate change the enormous coniferous forests that can magically summon clouds by emitting terpenes have to exert themselves far more vigorously. In northern latitudes change is happening particularly quickly, and the more intensely the sun burns in the sky, the more of these substances spruce and pine release to produce the cooling clouds. It is really astounding how much these forests have been able to do to help themselves – until now.

Trees can't react to human-caused changes in the short term, of course: their long lifespan makes that impossible. They live far too long to be able to do that. You can only have genetic variations over a succession of generations, and, depending on the species, these opportunities occur only every couple of hundred years – sometimes only after thousands of years – when the mother tree ends her life and makes space for her offspring. And when climate fluctuations within its lifespan become the norm rather than the exception, then the tree, or rather the whole forest, has to come up with a strategy to compensate.

To respond to the changing climate, trees have to be able to move from the spot where they are growing. Yet they aren't capable of doing that. This is a real dilemma, because each species is adapted to a particular climate where it can thrive. Whereas coconut palms need tropical temperatures year round and die a miserable death in the cold, deciduous trees can't sustain a growing season without taking a winter break. That's not a bad thing, you could argue: it means each species grows in the very place where climatic conditions suit them to perfection. And it's only because the earth features such a wide array of conditions that tens of thousands of species of deciduous and coniferous trees evolved in the first place.

Yet these climatic conditions are changing all the time, and as far as trees are concerned, they are changing relatively fast, even in Europe, where temperatures in recent centuries have fluctuated considerably. This happened in particular in the period known as the Little Ice Age, which some scientists believe was triggered by a series of volcanic eruptions.

After 1250, four fiery mountains near the Equator erupted, and their ash quickly entered the atmosphere and spread across the globe, blocking sunlight. As a result, so scientists from the University of Colorado Boulder believe, temperatures fell and glaciers expanded. The reflective properties of ice intensified the cooling effect, and temperatures sank even further. On average, it became 2.5 °C cooler – which is a lot when you consider the consequences that a global warming of 2 °C is predicted to have today. It wasn't until 1800 that things began to gradually warm up again. This was a very stressful time for trees, because each individual tree had to stay put and stoically endure the changes thrown at it. And it wasn't just cold all the time: some of the summers were extremely hot.[8]

Trees have only two strategies to survive this roller-coaster. First, most can survive in a wide range of climates. You can find beeches from Sicily to southern Sweden and birches from Lapland to Spain. Second, the genetic bandwidth within a species is very wide, which means that in any forest you can always find individual trees that deal with the new conditions better than most of their companions. In times of upheaval, these are the trees that reproduce and form new stands better adapted to the new norm. But for the extent of the fluctuations we're going through today, neither the beech's strategies nor those of the cloud-making conifers are sufficient. If it gets too hot, trees will get sick and soon be killed by bark beetles – weakened spruce and pines are their bread and butter, after all.

When it comes to the need to escape from high temperatures, it's all down to how fast the trees can travel. Does that mean that species with small seeds capable of flight have an advantage? Not necessarily, because trees have a big problem when it comes to reproduction. They have to provide their embryos, the seeds, with an energy reserve in the form of starch or oil and fat. In the first days of its life, the sprouting seed has to grow without being able to make any energy from photosynthesis. Roots penetrate the soil to get water and minerals, while above ground, cotyledons, or seed leaves, develop, which look very different from the later solar arrays that are so distinctive for each species. Only after these seed leaves have grown can water and carbon dioxide be transformed into sugar using light – and only then is the tiny sprout no longer dependent on the energy reserve bequeathed to it by its mother. And the size of this reserve differs for every species of tree.

Let's start with the smallest seeds, those of willows and poplars. They are so minuscule that you can just make out two tiny dark dots in the fluffy flight hairs. One of these seeds weighs a mere 0.0001 grams. With such a meagre energy reserve, a seedling can grow only 1–2 millimetres before it runs out of steam and has to rely on food it makes for itself using its young leaves. But that only works in places where there's no competition to threaten the tiny sprouts. Other plants casting shade on it would extinguish the new life immediately. And so, if a fluffy little seed package like this falls in a spruce or beech forest, the seed's life is over before it's even begun. That's why willows and poplars prefer settling in unoccupied territory.

You find the conditions they enjoy after a volcanic eruption, an earthquake, or a wildfire that completely obliterates plant life. In these landscapes, the tiny seeds can make good use of their advantages. Without rivals, they can grow up to a metre tall in their first year, and from then on, emerging non-woody plants and grasses can no longer inhibit their growth. The trick, of course, is to be the first to get to these places. Because the fluffy seed packages don't have an onboard computer, let alone any steering mechanism, the only way they can do this is by sheer numbers. A few of this fluffy multitude will always land in a suitable spot, and a mother tree of one of these pioneer species releases up to 26 million seeds every year. To keep the species going, every twenty to fifty years one of the little ones just needs to get a good start and reach an age when it, too, can reproduce. Does that sound wasteful? As trees have no idea where the ideal spots are, playing the numbers game is clearly the only way to get to the places they need to reach.

There are other ways of doing things, however. Just look at the beech and the jay working together. Flying is a good choice if you want to travel. Jays fly little more than half a mile before they deposit their plunder, but that's quite far enough for the beech. The tree's goal is not to reach those undisturbed spaces devoid of trees that are rare in central Europe, but simply to have the opportunity to travel in the first place. Trees need to be able to constantly expand their range a little to the north or south to follow climates that are always heating up or cooling down, even without the help of humans.

These changes usually happen so slowly that the restricted reach of birds is quite far enough. And for the beeches this is just one option for a small portion of their seeds, most of which will happily fall to the feet of the mother tree to sprout and grow in her shade. Beeches, as well as Douglas firs and other socially oriented species, love their families. If that sounds exaggerated, it's worth taking a moment to listen to the Canadian scientist Dr Suzanne Simard. She discovered that mother trees can sense through their roots whether the seedlings at their feet are their own children or the offspring of other trees of their own species. They support only their own children, by providing them with sugar through connected root systems – effectively by suckling them. But that's not all. To help the young trees, the parents step back underground, leaving the little ones more room, water and nutrients.

Where there are such family networks, what sense does it make to allow your offspring to be carried far away by the wind and by birds? Not much, and that's why beech nuts can't fly. Most seeds simply fall down from the branches and land in the soft leaf litter of the mother tree. Fast trips are not their thing.

However, if a beech nut happens to land in a spruce forest – because that's where a jay is setting up its winter stores – the seedling that sprouts has a good chance of surviving. It can handle low light, and it is patient. Millimetre by millimetre, it grows slowly upwards until eventually it reaches the canopy, where it can enjoy uninterrupted sunlight. Now it produces seeds of its own. Alone like this, far from its family, it must have a harder time than the other beeches, but it is fulfilling an important task. As soon as the temperature shifts a bit, it is the germ of a beech forest that will grow just a little further north.

Historically this has been a brilliant strategy, but right now trees with large seeds are moving too slowly. Should we help them? Couldn't we export beech nuts to Norway and Sweden and steal a march on establishing new beech forests, at the same time creating space for other trees, for example species from the Mediterranean region facing the same problem, which we could plant in forests in central Europe?

Apart from the fact that there are already beech trees in southern Sweden and southern Norway, I don't think this is a good idea. We know too little about how climate change will play out, and we don't know how local climates will develop. Warming doesn't mean that there will never be cold winters. It just means that cold winters will not happen as often as in the past. And if we import species of trees that love warmth, then an exceptionally cold winter might see them freeze to death. Aside from all this, a tree like the beech comes with a whole ecosystem containing thousands of species. We'd do better to concentrate our efforts on not allowing temperatures to rise too quickly – then the trees, with their slow rate of travel, will be all right.

There is, however, another kind of heat that can be even more dangerous for trees. And because some species of trees are pretty much like full tanks of petrol the situation can become too hot to handle.

13

It Doesn't Get Any Hotter Than This

A FOREST IS AN ENORMOUS storehouse of energy — its biomass, both living and dead, contains a great deal of carbon. Depending on the type of forest it is, it can amount to more than 100,000 tonnes per square kilometre, and if it burns it releases more than 367,000 tonnes of carbon dioxide (because of the two oxygen atoms that are added when wood burns). In coniferous forests, the trees contain dangerous flammable materials: sap and other easily combustible hydrocarbons. No wonder forests are always catching fire. Huge fires get going that often rage for months. Did nature make a mistake here? Why did evolution create species that are like open petrol canisters?

Deciduous trees, after all, show that there are other ways of doing things. As long as they're alive, they're absolutely immune to fire. This is something you can easily test for yourself — but please, with just a single green twig. No matter how long you hold a flame underneath it, the twig will not burn. Spruce, Pines & Co., by contrast, ignite easily even when they're fresh. But why?

Forest ecologists believe that in northern latitudes – the place most conifers call home – fire is a natural force for regeneration and even serves to preserve biodiversity. Under the headline 'Fire Creates Species Diversity', the website Waldwissen.net, which provides information on forestry for federal forestry administrators and professionals, published an article that sings fire's praises.[1] I find this an odd idea for many reasons. First, let's look at the concept of species diversity. To make a quantitatively verifiable statement on this topic, you have to know how many species are in the forests. But many organisms have not yet been discovered, even in central Europe, a region of the world that has been relatively thoroughly researched. And even if we know a species exists, we've often not done much research into how it lives and how widely it is distributed. Discovering a species doesn't say much except that it has been spotted somewhere at least once and we have its description on file.

There is a small beetle that lives in undisturbed forests which has been found behind the lodge where I live. In this particular region, it has only ever been seen in two other places, and these sightings both date back to the 1950s. Does this mean that this species is extremely rare? We don't know because, as in many other specialised fields, there's no money for further research. What we do know is that a weevil like the one found in the forest behind my home can survive only when conditions remain unchanged for a very long time. And because conditions in ancient forests rarely change much for hundreds of years, let alone thousands, the little beetles have lost their ability to fly. Why roam far afield if life is good close to home?

And so it's no surprise that populations of insects such as this one remain in the same place for a very long time. Their presence indicates that the forest has remained relatively undisturbed for centuries. A forest fire, probably extending over a vast area, would throw the whole system off balance. Where could the tiny inhabitants flee to? More importantly, how quickly could they run? A weevil could hardly escape a mighty wall of fire on foot. As far as I'm concerned, everything points to the fact that most forests in their natural state are not acquainted with fire.

There are other reasons why I find it odd to regard widespread forest fires as inherently natural phenomena. Humans have been playing with fire for hundreds of thousands of years – and depending on your definition of people, for much longer than that. If you include our forebears such as *Homo erectus* (Upright Man), then fire has accompanied our ancestors for about 1 million years. This was reported by researchers after they came across in Wonderwerk Cave in South Africa what were clearly the remains of cooking fires fuelled by twigs and grasses.[2] Examining the remains of teeth led to speculation that this relationship could go back twice as far,[3] and that modern humans developed their large brains because they enjoyed hot meals. Cooked food contains more energy and is easier to chew and digest than raw food.

No wonder that from then on people and fire became inseparable.

Fire, therefore, has not been an exclusively natural phenomenon for quite some time. Everywhere our ancestors lived, it was one of the very first by-products of incipient civilisation. So how can we distinguish between natural fires and those started by people? When it comes to forest fires in populated areas, it's impossible to distinguish between the two. How can we tell from charred layers today whether a forest fire was started by lightning or by a cave dweller creating a spark? You can't simply conclude that fire is part of a natural cycle in these places just because they happened regularly and forests always regenerated afterwards. The most you can say is that fire accompanies human settlement.

A strong argument against fire and forest naturally going hand in hand is the existence of individual trees that are extremely old. Take, for example, Old Tjikko, a spruce that stands in the Swedish province of Dalarna. According to scientific analysis, this puny little tree has at least 9,550 years under its bark, and it could get older still. If a forest fire had swept through the region in one of those years, Old Tjikko would have departed for the realm of its ancestors a long time ago.

Nonetheless, thousands of square kilometres of forest burn every year in Europe alone, above all in the south. There are many reasons for this. First, many ancient forests have been cleared. This dates back to the days when the Romans were cutting down trees to build their ships. Once the trees had gone, bushes took over, and forests could not be re-established because cattle, sheep and goats were pastured there, which meant that tree saplings had no chance of growing. To this day this shrub-covered landscape lies defenceless against the searing heat of the sun, and its dry bushes and grasses offer prime fuel for flames. In recent times, the forests that remained, often consisting of different kinds of oak, have mostly been replaced with pine and eucalyptus plantations. Unlike oaks, both these species burn like tinder, as the forest fire statistics of the last few decades clearly attest.

But the spark that sets off the wall of fire must come from somewhere. In rare cases, lightning is indeed the culprit. But for the most part, it's people who want the forests to burn, for a variety of reasons. Often, it's because they want to claim the land for development, which isn't allowed in forests in Europe. Once the trees have disappeared, new hotels and homes can spring up. This is what happened after the devastating fires of 2007. In Greece alone, more than 1,500 square kilometres of forest fell victim to flames, including almost 7.5 square kilometres in the Kaiafas lake reserve. But instead of allowing the area to regenerate naturally, the government decided to allow it to be developed for tourism and to retroactively approve about 800 buildings that had been erected there illegally.[4] Perhaps even worse are the motives of some firefighters. They put their lives on the line every time they go out to save people and property from the flames, which makes it all the more reprehensible that there are a tiny handful who, to ensure their jobs are secure, start fires themselves when things calm down too much.

Most fires can, directly or indirectly, be traced back to human activity. Even though flaming infernos don't usually have a natural cause, foresters still use them as a burning excuse for clear-felling: if clearing forests by fire is a natural occurrence, so the argument goes, such harvesting can't be detrimental. After all, nature itself creates open ground. Yet the opposite is true. Ancient deciduous forests in Europe had one important characteristic in common: long periods without change. And that's why the trees never developed any defence against fire. Despite their being extremely difficult to set fire to when they're alive, their bark doesn't tolerate heat. Beeches, for example, are so sensitive that if they grow in a clearing they get sunburn.

Even though forest fires are rare exceptions in most of the forests around the world, there are some ecosystems that are adapted to such events. Not to the complete incineration of all the trees – that would be an unforeseen catastrophe for any forest – but to fires that burn along the ground. These surface fires are another thing altogether, because they destroy only low-growing vegetation such as grass or non-woody plants but not the trees – at least, not the old trees. It turns out that old trees in fire-adapted ecosystems are equipped to withstand high temperatures periodically, and you can see this in their bark.

There is, for example, the coast redwood (*Sequoia sempervirens*), one of the mightiest trees in the world. It can grow more than 100 metres tall and live for many thousands of years. Its bark is soft, thick and slow to burn. If you find one of these in a city park – and you can find them in such places all over the world – step right up to it and press your thumb into the bark. You'll be surprised how soft it is. It holds a great deal of trapped air, which insulates the tree most effectively. Thanks to the insulating qualities of its bark, the trunk can survive unscathed a quickly moving wave of flames, such as those created by summer grass fires or fires in the undergrowth.

But it's only older individuals that can protect themselves this way. Redwood children have such thin bark that they are heavily damaged, often consumed altogether, by fire. Redwoods, therefore, expect to face fire over the course of their long lives, but they don't need it to survive: that's where the confusion lies. And, incidentally, it shows that even species which are adapted to fire don't like to burn. Quite the contrary, in fact. In places where fire is a natural component of the ecosystem, mature trees are designed not to combust easily precisely so that large areas are not reduced to smoke and ash.

The ponderosa pine (*Pinus ponderosa*), which is also native to western North America, is another tree that grows thick bark so that heat from forest fires doesn't damage its sensitive cambium. Ponderosa bark works like redwood bark: it protects older trees, and then only as long as the flames don't reach the crowns. This is where the needles grow, and they are filled with flammable substances. If the fire gets up there, then it quickly jumps from tree to tree, destroying whole forests. Again, the trees that are supposed to prove that fire is a natural phenomenon show only that even they actually abhor it. They've come up with a solution to rare lightning strikes and the surface fires they set only because they have the potential to live for a very long time: these defences allow them to survive to a ripe old age.

In my opinion, the supposed and much-vaunted benefits of releasing nutrients by flames and recycling dead biomass through fire are myths that downplay the disruption caused to sensitive ecosystems since prehistoric times by people playing with fire. In the normal course of events, it's not fire that releases stored nutrients and makes them available to new plant growth in the form of ash: it is the billion-strong army of animal sanitary engineers that undertakes the drudgery of decomposition, and which are themselves completely incinerated in large forest fires, because, unfortunately, these little fellows have very thin skins.

In the animal world, as in the human world, those who do menial labour toil away in obscurity. The many thousands of species that are small and unattractive are of little interest to humans. Beetle mites, anyone? They only make us think of dust mites, and even the thought of them gives us goose bumps. How about wood lice? When you find them under your doormat, you don't usually have much sympathy for them, either. The same goes for many other species that bustle about in the leaf litter under trees. But they're much more important for the ecosystem than, say, large mammals, because without these tiny, overlooked creatures the forest would suffocate on its own waste.

Beeches, oaks, spruce and pines produce new growth all the time, and have to get rid of the old. The most obvious change happens every autumn. The leaves have served their purpose: they are now worn out and riddled with insect damage. Before the trees bid them adieu, they pump waste products into them. You could say they are taking this opportunity to relieve themselves. Then they grow a layer of weak tissue to separate each leaf from the twig it's growing on, and the leaves tumble to the ground in the next breeze. The rustling leaves that now blanket the ground – and make such a satisfying scrunching sound when you scuffle through them – are basically tree toilet paper.

Whereas deciduous trees drop all their greenery at the same time and stand there stark naked, most conifers keep a few years' growth of needles on their branches and jettison only the oldest. This has to do with their native habitat. In the high north, the growing season is short: there are only a few weeks to grow leaves and drop them again. The tree would barely have time to be green before autumn came around and it would have to drop everything again. It would be able to photosynthesise for only a few days, and forming new growth or fruit would be almost impossible.

That's why Spruce & Co. retain most of their needles and store antifreeze for the winter instead, so that their needles don't freeze when temperatures drop. As soon as the first warm days arrive, the conifers can start producing sugar at full throttle, without having to expend energy and time leafing out. It's as though they are constantly on standby, ready to exploit the brief summer. However, thanks to their larger surface area, they are more likely to get blown over by storms or weighed down by snow. They reduce the risk by narrowing their crowns. In addition, because of the short growing season, they increase their height and girth extremely slowly, and it can take them decades to grow even a few metres. This means that storms get correspondingly little leverage, so the risks and advantages of being green all the time balance each other out.

In climate zones with clearly defined seasons, a tree's greenery must fall. Even in the tropics each leaf eventually serves out its time, and when it is used up and ragged, the tree replaces it with a new one. Inevitably, then, each such a solar panel, whether it comes from a deciduous tree, an evergreen or a conifer, drifts down to the ground. And there it would lie forever, buried under layers and layers of more fallen leaves many metres deep, until one fine day the ground would be depleted of nutrients and the forest would be full to the top with leaves – and then it would die.

And that's where the billion-strong army of bacteria, fungi, springtails, beetle mites and beetles come in. These tiny creatures are not trying to do the trees any favours. They are, quite simply, hungry. Each one goes about the business of processing its share of the bounty. One savours the thin layers between the leaf veins. The next enjoys the veins themselves. Others concentrate on breaking down the crumbly excrement of those who spearhead the attack. In central Europe, this group effort takes about three years. After multiple stages of processing, a leaf is finally transformed into pure faecal matter or, to put it more appetisingly, into humus. Trees can now send their roots out into this layer and use the nutrients that have been released in the decomposition process to build leaves, bark and wood.

But wait a moment: what happens to the substances the tiny little guys have eaten and incorporated into their own bodies? Well, the leaves' fate awaits them, too. In the best-case scenario, they are eaten once they are dead and their component parts are excreted. Under less happy circumstances, their end comes more quickly, while they are still alive. Small dramas play out in the leaf litter every day. Just as lions on the savanna hunt and eat gazelles, so spiders and beetles in the forest hunt and eat springtails. Hundreds of thousands of tiny creatures and many hundreds of hunters are to be found in every square metre of forest floor that is covered with a thick layer of humus. If you have good eyes and a lot of patience, you can watch the activity for yourself. Depending on the species, springtails can be a few millimetres long, and spiders and beetles are even bigger.

The substances gathered up in the animals soon get back into circulation when they are excreted, and then become available to plants as well. There is just one thing the tiny creatures don't like, and that's cold. When it gets too cold, they stop work. And it does get cold in the deeper soil layers 10–20 centimetres below the surface in an intact forest. Humus washed down to these depths by the rain remains basically untouched, even by fungi and bacteria.

Over thousands of years, this blackish, brownish layer increases in depth, and sometimes, because of geological processes, it forms coal. Other material is washed deeper and deeper or, it would be better to say, seeps along many floors down with an extremely slow flow of water over the course of decades. And down there, the unhurried underground dwellers we met earlier are waiting. The deeper they are, the less time seems to matter to them. They, too, prefer organic substances to ash, which brings us back to forest fires: all this shows that nature has come up with a much smarter and less incendiary way of recycling nutrients, one in which thousands of species benefit, instead of being burned to a crisp.

These natural recycling systems, however, are mostly no longer functioning as originally intended. People are manipulating them and interfering with them in many different ways, and not just by playing with fire.

14

Our Role in Nature

LET'S CUT TO THE CHASE and start with one of the biggest difficulties here, namely the answer to the question: What exactly is nature? Do we mean by it untouched tropical forests and remote mountains with unscaled peaks? Or is it flower-filled alpine meadows grazed by dun-coloured cows with large bells swinging from their necks? What about abandoned strip mines, where pools have formed and frogs now croak loudly? There are probably as many definitions out there as there are people who love nature. One simple standard definition is that nature is the opposite of culture – everything that people have not created or changed. This definition draws hard and fast boundaries around what can be called nature. Other definitions see people and their activities as part of nature. From this perspective, nature and culture cannot be clearly separated.

And right here is the problem faced by the modern conservation movement: What is truly worth protecting in nature, and what counts as a threat or even a disturbance? When the nature you're thinking of is close to home these can be tricky questions to answer definitively.

However, as soon as you let your gaze wander further afield, the situation looks quite different. Of course the Amazon rainforest should remain as intact and undisturbed as possible. And please leave Antarctica – a place that under international law doesn't belong to any country – untouched! You'll find similar sentiments when it comes to coral reefs in Australia, for example, or ancient forests in Kamchatka. In your own backyard, however, a much more malleable rule applies, which states that under some circumstances culturally manipulated landscapes are also worthy of protection, particularly when the original landscape has completely disappeared.

I tend to side with those who argue for a clear separation, otherwise one day palm-oil plantations in Borneo will count as part of nature, too. But how easy is it to make this distinction? Which historical epoch signals the divide after which people should be counted as a disruptive influence? If we tend to see our species as a disruptor ever since it arose, what about our predecessors, such as *Homo erectus*, who differed from us only slightly? There are many questions for which we have no clear answers. I like to draw the line with the beginning of agriculture. As soon as hunters and gatherers settled down, selective farming practices began to change species. This was also the time when the landscape began to be deliberately transformed into an ecosystem completely devoted to meeting human needs.

This was when the first irreversible disruptions of the environment became visible, for example as a result of ploughing. When they are drawn over the land, ploughs disrupt layers deep in the soil profile. The soil retains the imprint of these so-called plough pans for tens of thousands of years. Water drains poorly, and even oxygen has a hard time penetrating the barrier they form. As a result, the roots of many tree species rot when they try to grow down through them, and the trees grow wide, shallow root systems instead. This makes them unstable, and when they reach a certain height (usually about 25 metres), the leverage of storms is so great that they topple over.

Just like the birds or bears we've looked at, we too influence the forest and the kinds of trees that grow there, and not only as a result of accidental changes caused by our agricultural practices. Today, 98 per cent of the forested areas in Germany are planted, cared for and harvested on an industrial scale, but as far back as the Stone Age our ancestors, who were wandering around not with ploughs and saws but with bows and arrows, were already managing to do a fine job of disrupting nature. I'd like to take you back into the past with me, to see what our ancestors wrought several thousand years ago with the modest means at their disposal.

At the end of the last Ice Age a dramatic change in climate occurred. About 12,000 years ago the remnants of glaciers a kilometre thick finally melted, exposing a desolate landscape. There were no forests left in central Europe: they had all been destroyed as the glaciers slowly advanced from the north. The trees had nowhere to go, because the glaciers in the Alps also advanced, blocking their path like a gigantic girder across the landscape and preventing them from escaping south. Many species died out; others were reduced to a few remnant stands in ice-free side valleys or survived only in the warmer climes of southern Europe.

When the ice melted, the vegetation cautiously returned. At first, there were just mosses, lichens and grasses, quickly to be joined by miniature bushes and trees. A tundra developed, like the ones we find today in the northern regions of Canada, Scandinavia and Russia, where you can still see what a post-ice age landscape looks like. Later, the trees returned. First, it was conifers such as pines that, along with birches, could best withstand the cold that still reigned in these parts. As time went on, oaks and other deciduous trees joined them, pushing the conifers out of most habitats once again.

One representative of the conifer class, however, seems to have dragged its feet: the silver fir (*Abies alba*). It moves very slowly, and so far it has only made it as far as central Germany. You can see the sequence of the trees' return in the Alps today, by the way. High up on the slopes, where the Ice Age still dominates, you find glaciers. The lower you descend, the warmer it becomes, and the more plants you find – and those further down on the slopes are getting bigger, too. Four thousand to five thousand years ago, beeches also returned from the south to central Europe. Today, they would form the vast majority of our forests if – and it's a big if – modern humans had not continually interfered by cutting them down and planting other species.

But is it really only modern humans who are responsible? After all, our forebears returned along with the plants when the ice retreated, after their ancestors had also been forced by the glaciers to relocate to southern climes. These returnees were far too few in number to be able to damage the nascent forests. Within the borders of what is Germany today, there were perhaps no more than 4,000 people roaming around the barren landscape. Then, along with further warming and reforestation, the human population rose, and by 4000 BCE, there were some 40,000 people around. Still, that was not more than one person every 100 square kilometres. Even if these people needed to burn a lot of fuel, that wouldn't have had much impact on the forest, which grows more than 100,000 cubic metres of new wood annually in an area this size – about the amount of energy used by 1,000 modern single-family homes in Germany today.

The problem, then, wouldn't have been that Stone Age people were cold, but perhaps that they were hungry. They hunted large herbivores, and large herbivores like to eat young trees. The largest of these herbivores were aurochs and wisent (aka bison), as well as horses and rhinoceros. These species all specialise in eating grass, and they graze grassy plains so thoroughly that they stop any reforestation from happening. And this is of critical importance for the discussion we're about to have. If these animals, which naturally shaped their habitat, were present in high enough numbers, then the northern latitudes were probably not forested at all back then.

In the absence of forests, therefore, the secret rulers of the ancient landscape were not trees but large herbivores. Herds of these grazing aurochs, wisents, wild horses and deer roamed the grassy plains, demolishing every tree as soon as it emerged. At least, that's the theory. And even if, despite the herds, enough trees managed to get established to form a proper forest stretching to the horizon, horses and deer would have wasted no time stripping the bark off oaks and beeches, thereby killing them, and the trees' offspring would have been constantly trimmed by hungry herds biting off their buds and new shoots.

But of course all these large herbivores – except for the deer – have long since disappeared. Were they really wiped out by human hunters? Could a few representatives of the species *Homo sapiens* really have had such a powerful impact? Here's where Sander van der Kaars' international team of researchers comes in. They searched the coastal waters of Australia for traces of excrement left by extinct species of animals, and have concluded that human hunters who settled the continent about 50,000 years ago were responsible for the extinctions. They dismissed fluctuations in climate as the cause, because these were not as severe as they were in the northern hemisphere at this time. Less than 1,000 years after the arrival of the first Australians, 85 per cent of the continent's megafauna – animals with a body weight of more than 44 kilograms – had disappeared.

Yet the disappearance had nothing to do with excessive hunting. Quite the opposite. In the researchers' opinion, the large animals reproduced so slowly that even a moderate level of hunting inflicted grave damage. With each individual hunter removing just a single adult animal every ten years, the scientists calculated, that would have been enough to wipe out the species in a few hundred years.[1]

If large herds of wild cattle, rhinoceros, elephants and horses really did shape the landscape in central Europe before hunting humans began to interfere, then in a best-case scenario shrubby growth could have developed, but not endless tracts of forest. Of course, supporters of what is known as the 'megaherbivore theory' know that at that time central Europe used to be almost completely forested. But in their opinion this was because of people. Farmers in the Neolithic, they argue, hunted large herbivores intensively and decimated their populations, giving the forest an opportunity unforeseen by nature, and the forest seized its chance. They support this idea with pollen finds that confirm the presence of grassy plains vegetation prior to this time.[2]

However, there's also evidence of a massive amount of pollen from trees from the same time period. This doesn't have to contradict anything, because even in enormous ancient forests there would always have been places that were free of trees. Such areas could have been swamps, steep slopes or riparian zones where raging floods didn't give trees the chance to survive for long. The only question is how large these areas of grassland were. Did they dominate, or were they merely marginal?

There is one more argument in favour of treeless zones. Aurochs, wisent and deer are all herd animals. And herd life is only possible on the plains. Have you ever gone walking off the trail in a dense forest with a large group of ramblers? Then you'll know that the members of the group soon spread out and lose contact with one another. You have to keep stopping to wait for stragglers, and because you can't see them you don't know when they will show up.

For wild cattle, the situation is even more dangerous, because a herd attracts much more attention from predators than single animals do. There are the calls the animals use to communicate with each other; the enormous, strongly scented trail they leave; and, most importantly, the slower speed of the whole group as it has to stop and wait for laggards. For wolves and bears, it's tantamount to an invitation to an all-you-can-eat buffet.

Typical forest animals such as roe deer and their enemy the lynx roam completely alone. It is only at mating time and when they're raising their offspring that you find small family groups of two or three animals. There are also differences in flight behaviour. Whereas herd animals often run for kilometres before they stop, solitary forest animals usually move less than 100 metres. By then they're hidden in dense undergrowth and can calmly wait and see if their predator is bothering to pursue them or not.

So, just to recap, the remains of pollen indicate the presence of forest-free areas, a finding supported by the existence of large herbivores roaming in herds. It is possible that hunting by humans could have led to a drastic decline in their numbers, and the forest could have then reclaimed the now-empty plains. This theory is supported by the fact that most of the large and very large herbivores are now extinct. Mammoths, woolly rhinos, wood elephants and wild horses, aurochs and wisent – none of them exists today (except for a few animals in Białowieża National Park in Poland). And the warming trend of the past few thousand years is no sufficient reason for their demise.

So far, so good. But this theory is nonetheless rather shaky. Consider the situation from the other side: let's leave the herbivores and look at the trees. Forest trees native to central Europe, such as oaks and beeches, have completed a long selection process over many generations to become the rulers of ancient forests. A whole range of amazing abilities has allowed them to survive for many millions of years. But there's one thing these trees never developed: protective adaptations against large herbivores. No toxins, no thorns or prickles. Young trees in particular have no way of fending off the maws of deer, horses and cattle. If the megaherbivore theory were true, it would mean that deciduous trees native to central Europe lived under constant threat with no way to defend themselves.

OK, some deciduous trees, as we've seen, can identify roe deer and load up with defensive substances when the deer are eating them, but this defence doesn't help much if there is a high density of wild game out there. We know this from the futile efforts of forest owners to protect their trees. All the small beeches and oaks get nibbled so extensively that the damage stunts their growth for decades as though they'd been bonsaied, and even when there are too many herbivores around and food is scarce in winter and chemicals are sprayed on the buds to stop them from being munched, they get eaten too! Deciduous trees appear to be so delicious that when there's a certain density of deer, there's no way to save them.

These depredations don't happen to typical plains plants such as blackthorn and hawthorn – their names betray their defensive strategies. Even plants like stinging nettles and thistles have armed themselves. Pointed, hollow needles filled with toxins; bristles that break off easily and remain attached to skin; and tough, bitter fibres – these are among the methods plants use to keep greedy browsers away. On top of this, they can disperse their seeds via airmail on the wind or by bird so they can quickly settle available spaces, even if they're some way away. Beeches and oaks, by contrast, stand there totally defenceless. As I've said, their heavy seeds are dropped right at the feet of mother trees, and travel no more than a couple of miles even when animals carry them away. It takes these trees thousands of years to move to unforested areas.

The only reliable conclusion we can draw from all of this is that there was never a substantial threat from grazing herds. Corroborating this inference is that it takes a native forest in central Europe about 500 years to achieve a stable balance. Millions of hungry hoofed animals would never have allowed the trees so much time. The bottom line is that despite evidence of plains plants and large herbivores, forests must still have dominated the region. Even supporters of the megaherbivore theory concede that there were beeches and oaks around. If these had been isolated patches, they would quickly have been stripped bare. Moreover, their seeds are so heavy that they would not have travelled hundreds of miles on the wind but only short distances with the help of birds. That these defenceless species of tree were still to be found all over the place speaks against herds of landscape-altering horses and cattle.

This conclusion is bad news for foresters and hunters who misuse the megaherbivore theory for their own ends. Foresters see no problem with clearings regardless of whether they are made by browsing aurochs or lumberjacks. While hunters use feeding stations to artificially increase the number of deer that then devour every small deciduous tree for miles. Both parties rely on the megaherbivore theory to argue that, based on historical precedent, neither open spaces nor heavy browsing is detrimental to the long-term health of forests. Hubert Weiger, the president of the conservation organisation BUND (Friends of the Earth Germany) in Bavaria, has warned: 'An intellectually interesting and highly technical discussion about conservation... is being exploited by certain land users... as a political tool to get their damaging objectives implemented.'[3]

Apart from natural fluctuations in climate, forests now have to contend with the disruptive climate change we have set in motion. Change is happening quickly – too quickly for trees. In the summer of 2016, returning at the end of August from my summer holiday in Norway, I noticed a strange phenomenon, and what I saw frightened me. When we travelled to Scandinavia, we had left the forest I manage in a healthy state of green. During our week-long absence I wasn't concerned. At Hardangerfjord, our destination, it rained so much that I hankered for the weather that was being reported for Hümmel: bright sun and temperatures higher than 30 °C. When we got back, and after a long drive finally caught sight of our native beech forest, I didn't feel quite so happy. Over the course of those few hot days many of the crowns had turned brown, and some of the trees were already missing most of their leaves.

I soon convinced myself that it couldn't have anything to do with lack of water. I took a few soil samples from different places and pressed the cores between my thumb and forefinger. The soil didn't crumble: I could press it into little discs that held their shape, a sign that there was enough moisture. So what was up?

When trees shed their leaves in summer, it almost always has to do with agonising thirst. They would rather discard their leaves – the surfaces where they lose the most water – before they dry out completely. Unfortunately, such drastic action ends the season for them, as they can't photosynthesise any more. They have enough energy left to grow new leaves the following spring, but not much else. A late frost that freezes the fresh leaves and forces the tree to start over, an attack of insects during which reserves are expended in producing defensive compounds – stresses like these can exhaust beeches and oaks to the extent that they die. With spruce, death comes in an even more spectacular fashion. Their needles turn a fiery red and, because the dying tree is soon discovered by bark beetles and attacked, not only do the branches lose all their needles but bark also falls off, exposing the trunk.

But back to the summer of 2016. Until August, it had been cool and damp in our area, conditions trees usually love. Usually. In central European latitudes, too much rain in summer can favour pests. In July these had been the cause of an initial leaf drop. At this early date, fungi had been enjoying a feeding frenzy in the leaves, sprinkling them with brown spots or covering them with a thin milky layer of mildew. When their green solar cells were overwhelmed, the trees got rid of them. Some days, leaves were dropping from the crowns as though it was already autumn. And then came the quick, fateful change to extremely hot, dry weather. Such a sudden switch is enough to throw even the strongest trees off balance. Within a few days, many of the deciduous trees in my forest had turned brown and finally discarded the leaves the fungi had left undamaged.

It was notable that in the managed stands, where trees are repeatedly felled, the symptoms were particularly prevalent. That's not surprising, because here, in contrast to forests left to their own devices, there are many gaps in the canopy through which the sun can shine unchecked. This means that everything heats up more quickly, the air dries out just as rapidly, and all the conditions change much more abruptly. Conversely, in forests that are left alone to regulate their microclimate for themselves, conditions remain somewhat more bearable. Additionally, the trees in these forests mutually support each other through their root and fungal networks so that weakened friends can be saved.

And what about weather conditions at other times of the year? As a forester, I've always got one eye on the weather. When it's stormy in winter I worry about the old spruce trees that might fall. If they collapse, the small beeches growing under them, which still need the shade of their guardians (even though they are not related), will be exposed to the sun the following summer with no way of protecting themselves. If it rains too much, there's a greater risk that the ground will soften and not offer the roots much in the way of support. I prefer freezing cold days in winter, because that means there won't be any precipitation. For it only gets really cold under a high-pressure system, when cloudless night skies allow warmth from the earth to radiate into outer space.

But if there isn't any rain or snow, that's not good either. In summer, trees in central Europe don't get enough water from rain clouds, so they have to draw on reserves that have built up in the ground over the winter. A lot of moisture is stored when the trees are not actively growing, and they can then use this moisture to supplement the rain that falls in the warm months of the year – as long as there was enough precipitation the previous winter.

Hot summer days also worry me. Too many in succession, and the ground dries out and trees suffer. They become more susceptible to disease, as I've already explained. If rain comes, it's often accompanied by a thunderstorm. Right before the storm hits, the wind freshens sometimes to storm strength, and the deciduous trees I love so much are now particularly at risk, because they present a large surface for the wind to attack. In winter, on the other hand, which is the usual time for storms in Europe, they present a streamlined, leafless profile, as evolution has taught them to do. So I don't like summer thunderstorms.

You'll have noticed by now that the weather gods just can't satisfy a forester like me. By way of apology, let me say that I am simply worried about my trees and their future. Because I'm paying attention every day, I notice subtle changes to our climate year on year. It's not just the mild winters, which are being discussed all over the media. There's also a noticeable shift in the seasons. The first snow often doesn't fall until January, even though the forest I manage, which stands at an elevation of 500 metres, should normally get at least one sprinkling of white by November at the latest. And then March often slips by without offering any days warm enough for me to sit outside.

The bees are being left out in the cold, because either meadow flowers and other early sources of nectar are late appearing or low temperatures prevent the insects from going out on foraging flights. And while garden centres are offering a wealth of flowers for windowboxes and flowerbeds, we have to wait until the middle of May to fill the garden outside our forest lodge with colour. The last snow of the season falls in April, and sometimes there's still a spell of frost in early June, so the petunias that were purchased prematurely have to be replaced. In Germany lately, it hasn't been getting really hot until August, and in 2016 it didn't heat up until the middle of September. From a meteorological perspective, autumn should have arrived by then – with a few final warm days to delight us, sure, but temperatures should have been distinctly cooler, especially at night.

All things being equal, we might be fine with delaying a little the things we do at certain times of year, but unfortunately, trees' internal clocks are set differently, or perhaps they are simply more stubborn. They're as acutely aware as we are that the days are getting shorter, and they slowly prepare for winter dormancy. Simply holding on to their leaves for another four weeks isn't an option for them, however, because despite the warming trend, they still have to reckon with the possibility of an early onset of winter accompanied by a heavy snowfall, which would punish trees that keep their leaves for too long in the autumn sunshine. Their branches would snap and some trees would lose their balance and topple over, which is what happened in a big snowstorm in Germany in October 2015.

The only recourse trees have is to escape north, and that's what they are actually doing – or to be more accurate, *trying* to do. It never occurred to us that trees might migrate, so the parcels of forest land we have set aside to mark property ownership create inflexible obstacles for trees wanting to extend their range to cooler climes. Take our lawn, for example. When I mow mine, I always see small oak seedlings poking up out of the grass, which then unfortunately fall victim to my mower. OK, so the mother oak stands just 30 metres away, but even so, this is a migration, even if it's a very slow one. I've already explained how far birds and the wind can transport seeds, but if we have other plans for every patch of ground where seeds might land, the trees can't even get started on their journey north.

With animal migration, there are international efforts to keep corridors open so that enormous herds of gnu, zebra and elephant can move from one national park to the next. Even in central Europe there is support for animal migration – for example, to help wild cats. Conservation organisations like BUND strive to maintain corridors so that they can expand their range once again and roam everywhere in Germany.[*]

But what about trees? They advance so sedately that tree migration is not on anyone's radar. Even foresters say that Beeches & Co. are too slow to escape to higher latitudes as the climate changes. Yet the problem here is not that they are too slow, but that we are keeping their populations confined. Every time one of their seeds sprouts somewhere we've not designated as an area for trees, we immediately remove it. Spruce have to grow in district X, and beeches have to grow in district Y. A parcel of land over there is to be used for agriculture; another is listed as a meadow. These rigid boundaries get in the way of what nature has in mind: change. And this brings us back to my lawn – and, yes, I too am guilty. If we've force our environment into a straitjacket, how can we have any idea how it reacts to climate change? Are trees in central Europe really too slow to set out for the cooler climes of the North?

Apart from protecting the climate in general by conserving energy, of course, I think the solution lies in designating many more protected areas. We need zones of wild forests to be like the stepping-stones we use to cross a river without getting our feet wet. If there were enough of them, wild species could travel freely through our culturally manipulated landscape from one reserve to the next. And if these areas were not too far apart, then perhaps we really could see how trees react to climate change, and we might even find out that they don't want to go north at all.

We already know that as long as beech forests are not disturbed by commercial forest practices, they can cool themselves in hot summers. It's only when trees are felled and sunlight penetrates the shade under the remaining dark trunks, and the air down there dries out and heats up, that they start to have problems. That points to a simple and almost banal solution: less use of wood equals less use of energy equals slower climate change equals more healthy, resilient forests. If this could happen in certain areas, then there's hope for the slow-moving giants of the plant kingdom.

Some effects of human activity on nature are much more subtle and difficult to track than felling trees, simply because cause and effect lie much further apart.

In 1997 I travelled through the south-western United States with my family for the first time, and twenty years later I made a return visit. We were blown away by what we saw in the United States. The national parks, with their imposing sandstone rock formations, are simply breathtaking. Apart from the plants and animals in that vast landscape devoid of people, it was the bizarre shapes of the rocks in Utah that particularly grabbed our attention.

Arches National Park is named for the unusually high number of impressive rock arches it contains. Some of these gigantic monuments look so fragile, despite their imposing size, that amazed visitors wonder how they've remained standing under wind and weather for thousands of years. Today, for many of them, this question no longer needs to be asked. Since 1977 forty-three arches have tumbled in Canyonlands National Park alone.

Several of these tragedies – for that's what they are, not only for tourists but even more so for the indigenous peoples for whom the arches are sacred – can be traced back to human activity.

A research team at the University of Utah has discovered that the rocks sway ever so slightly, for a variety of reasons. Most of the movements are caused by natural events. After earthquakes, the main culprits are temperature fluctuations. The rock expands in the heat of the day, and then contracts as it cools down at night, which makes the arches sink slightly into the ground.

To get to the bottom of what else was causing it, the scientists wired up Rainbow Bridge, which is considered to be one of the highest natural bridges in the world and remains a sacred site for the Navajo Nation. Tourists are not allowed near it. Those who want to see it have to travel by boat along a side arm of Lake Powell, and from there a ranger walks them to an observation area. These precautions have less to do with protecting the arch than with respecting the beliefs of indigenous people who live there. But tourism is also not the main threat to the bridge.

As Jeffrey R. Moore's team of scientists discovered, the repercussions of human activity are detectable in the rock – in fact, they can be felt every few seconds. The rhythm of waves gently hitting the shore of Lake Powell can be measured on Rainbow Bridge many miles away, where the wave action causes small but continuously repeating vibrations in the rock.[5] If such gentle motion is measurable, it's no surprise that shock waves from drilling for oil and gas in Oklahoma – a distant 1,000 miles away – were also detected. Ultimately, it's difficult to say precisely what has led to the collapse of arches in the recent past, but the research demonstrates clearly the effects human activity has on ecosystems.

Forgive me for going off on a tangent, but the problem of the collapsing arches I've just been explaining gave me an idea. It's pure speculation because, as far as I know, it's never been tested. It has to do with groundwater, and goes like this. Water deep underground contains gas. This gas includes the oxygen that groundwater crustaceans and other tiny creatures need to breathe and, it follows, the carbon dioxide they exhale. You know what happens when you shake a bottle of carbonated water: the carbon dioxide bubbles out and the water then contains less of this gas and becomes less acidic.

You could therefore compare the groundwater system to an enormous bottle that is constantly being shaken by artificially induced tremors. Surely that must result in changes in the gas and acid content of the water? It could be the case, at least in areas around fracking sites where pressurised liquid is used to fracture the ground up to 3,000 metres below the surface, a process that produces numerous earthquakes. As collateral damage, many chemicals remain in the ground, their fine particles dispersing and infiltrating cracks in the layers that are being worked. What would the blind crustaceans have to say about all these changes to their ecosystem? I wonder.

*

In central Europe the majority of the subterranean rivers that form this wonderful groundwater ecosystem are still untouched, but there have been dramatic changes close to urban developments. For one thing, pollutants from agriculture and industry are seeping underground. For another, enormous quantities of water are being pumped out of the ground. In Germany alone, about 10 million cubic litres of water pour out of taps every day. And then there are industrial uses such as open-cast mines, which are emptied of huge amounts of groundwater that flow into them. At the open-cast soft coal mines near Cologne, 550 million cubic metres of water was pumped out in 2004 alone – that's one and a half times the drinking water used in all of Germany in a whole year. An underground area of at least 3,000 square kilometres is affected, where life-forms that we haven't studied yet and whose influence on the natural cycles of life we don't yet know are wriggling around in every cubic metre.

Despite this, there are still large regions where the groundwater system remains intact, and together with the deep layers of soil, these really are the last untouched habitats in central Europe. So you don't have to go far to find true, untouched nature: it's closer than your closest national park or conservation area, but still beyond your reach.

What is both close by and immediately accessible, however, is the results of the last 100,000 years of human evolution.

15

The Stranger in Our Genes

HOMO SAPIENS HAS TURNED OUT to be a very successful species (thus far), which is probably an indication of our aggressive nature (I will explain exactly why in the course of this chapter). I don't mean our urge to attack each other, but our tendency to attack other species. This aggression has something to do with our evolutionary success, which has made us what we are today. And if the decline of other species is anything to go by, perhaps we've been too successful. Is the desire to disrupt the giant equilibrium of nature stored somewhere in our genes? Or have we succeeded in removing ourselves from her clockwork mechanism and entered into some kind of parallel ecosystem?

I often hear people talking about how modern humans have halted evolution in its tracks. This notion is linked to medical advances. How many of us would still be alive without appendix operations, insulin injections, beta-blockers, or even spectacles? Ten thousand years ago, the defects that plague us today would have made us an easy target for predators. The hard truth is that evolution would have weeded us out. So if we survive with medical help despite our physical weaknesses, and if we then pass our defects on to the next generation, does that mean our species is becoming more fragile and would die out if medical help were to suddenly disappear?

To take a closer look at this, we first have to differentiate between two separate questions: first, whether evolution has actually been switched off; and second, whether the use of medical aids might itself be part of evolution and indicate an advanced stage of development.

The answer to the first question is clear. Evolution is obviously still hard at work, particularly when it comes to disease. For many people, the levels of danger and pressure exerted on them by disease in their environment are as intense as they have ever been. According to the World Health Organisation, in 2015 alone 200 million people fell sick with malaria, and 440,000 of them died from the disease. In places where malaria is widespread, a rare genetic blood disease is also prevalent: sickle-cell anaemia. In sickle-cell anaemia, red blood cells, which are normally round like a disc, become sickle-shaped. People who suffer from sickle-cell anaemia have difficulty getting enough oxygen to their organs, and they often die before the age of thirty. Most people who carry the genes for the disease, however, develop only a mild case, which means that although they have sickle-shaped blood cells, they also have enough normal-shaped cells to lead almost normal lives.

Crucially, however, the disease offers some protection against malaria. In this illness, parasites passed on through mosquito bites attack and destroy red blood cells. Malaria advances in stages. Periods of fever, triggered by blood cells bursting en masse, often lead to the complete breakdown of the organism. Yet carriers of sickle-cell anaemia have a natural resistance to malaria. How this works has yet to be adequately explained. In any event, people who have sickle-cell anaemia and are significantly affected by this disease have a distinct advantage over those who don't. This advantage means that in areas where malaria is widespread, you also find many people with this genetic mutation.

As this example demonstrates, the impression that evolution has almost ceased and that humans have achieved the pinnacle of their success is false. It's just that people in wealthy industrialised countries have become somewhat removed from the processes going on around them. Nature continues to apply pressure. Cancer, heart attacks and strokes are only some of the factors we cannot control despite medical advances. Strictly speaking, it's modern civilisation that makes modern medicine necessary in the first place. The afflictions that are so aptly named 'diseases of civilisation' barely existed thousands of years ago. Dental braces, back operations and heart bypass surgery are only necessary because of unhealthy Westernised lifestyles. Seen in this light, medical discoveries that have supposedly halted the rollercoaster of evolution have simply pushed it in a different direction. Instead of plagues,

Cholesterol & Co. are now the factors sorting out the gene pool.

Apart from that, numerous building sites in our bodies are proof of archaic developmental processes that still function as they used to. Our mouths are losing unnecessary teeth (wisdom teeth), our guts are losing unnecessary appendages (the appendix), and our bodies – to the intense regret of many men – are losing their hair. It's unlikely that people in 50,000 years' time will look exactly the same as they do today. So evolution is going along its merry way, even if we are under the impression that we've reached the end of a long journey. It's just that change happens so slowly that we don't notice.

Something similar happens to the surface of our planet. The appearance of the land masses and the shape of the continents seem to be unchanging, even though in school we all learned about the movement of the tectonic plates that form the earth's crust. These plates, which span whole continents, drift on molten rock either towards each other (which thrusts up mountains) or away from each other (which opens up rifts where lava gushes to the surface). North America and Europe, which lie on different plates, are drifting further apart by about 2 centimetres a year – about twice as fast as your fingernails are growing. Apart from a handful of scientists, hardly anyone takes any notice, but over 10 million years – just a brief moment geologically speaking – that amounts to 200 kilometres. It's only when things occasionally get stuck and the jammed plates rip free again that earthquakes alert us to the upheaval.

This raises an important question. Does evolution move at different speeds and in different directions in different places? Whereas some people feel the full impact of the selection process in the form of disease, for others – mostly in industrialised countries – the impact is softened considerably by a wealth of resources. However, what might seem like an advantage for individuals could, in the long term, be a disadvantage for the population as a whole in any given place. In the West, victory over plagues has shut down two of the most important factors that have constantly changed our genetic make-up. If for people in wealthy countries evolution has indeed almost come to a standstill, over many thousands of years they could still be overtaken, genetically speaking, by people who live in countries with less access to medical advances and Westernised lifestyles.

In practice, however, such developments are impossible in our modern world, because our enormous mobility interrupts this process of separation. Migration increasingly blurs local differences. Many people today have ancestors who lived in other countries, and so a gradual genetic drifting apart of the people of the earth is no longer on the cards. This also means that the development of different human species is out of the question – at least for now. For that you'd need populations to be isolated for a long time, something that in an age of world travel and migration has become impossible. Scientists tell us that every person alive today can be traced back to one mitochondrial Eve, who is said to have lived 200,000 to 300,000 years ago. The variations in skin colour and other characteristics that have developed since then are disappearing increasingly quickly. What some people mourn as a loss of diversity, others embrace as an opportunity for humanity to bid goodbye to racial differences.

Evolution, however, can advance in totally unexpected directions. There was a time when *Homo sapiens* was not the only hominin species to walk this planet. To explore this further, let's turn to some of our dusty distant relatives from the Neander Valley in Germany. Heavily muscled Neanderthals from the Stone Age had a brain comparable in size to ours. Their culture was relatively advanced. There was a division of labour in their settlements. They crafted ornate stone spear tips attached to wooden shafts, painted their bodies, buried their dead, and spoke in a language that has long since fallen silent.

Scientists assume that *Homo sapiens* and Neanderthals lived side by side in Europe for several thousand years. Could it be that the modern humans who arrived later on the scene than the Neanderthals learned something from watching their stockier neighbours? Could it even be that this species of human was at the time mentally equal to the *Homo sapiens*? Scientists debate this question, but I don't think they do so fairly. Earlier *Homo sapiens* were no different from people today, so if you agree that Neanderthals and *Homo sapiens* were mentally equal, you're saying that the intellectual prize we thought was ours alone has to be shared with another species. And it appears that evolution handed the crown to humans not because of the size of their brain but because of their more aggressive tendencies – after all, we supplanted the Neanderthals and might even have used them as a source of meat.[1]

There are some arguments against this interpretation, but an objective discussion isn't yet possible, because people are still attributing to Neanderthals only those minimal intellectual capabilities that can be proven by archaeological findings. Take language, for example. Neanderthals had a small bone under the tongue, the hyoid. This bone is a necessary requirement for speech. There is also a particular gene, FOXP2, that is considered indispensable for understanding verbal utterances, and Neanderthals had this, too. But for scientists, this isn't proof that Neanderthals used language: all it proves is that they were physically capable of doing so. If you take that tack, the presence of eyeball sockets in unearthed Neanderthal skulls only means that they had eyes. But whether they could actually see with them can never be said with absolute certainty.

The size of the Neanderthal brain has been explained away as either an adaptation to the cold or to their slightly heavier bodies. (The same sized-brain in a larger body means that proportionally the brain is smaller.) However, there are people alive today whose weight and musculature are pushing the Neanderthal range and yet their brains are 'only' Neanderthal-sized.

If that argument held water, body builders would be packing on brain cells as well as muscle while they were at the gym.

A few years ago an important tenet of science fell by the wayside. It stated that Neanderthals and modern humans did not interbreed; therefore, no material from our coarser cousins can be found in our genes. Yet mapping the human genome turned up a number of surprises that gave us a fresh glimpse of the Neanderthals. Today, researchers believe that 1.5 to 4 per cent of the genetic heritage of most people of European and Asian ancestry comes, somehow, from Neanderthals.[2]

Our extinct relatives greet us through the skin and eye colour of many of our contemporaries. A light complexion and blue irises, for example: current scientific opinion holds that these are Neanderthal adaptations to their northern habitat. Here the sun is less intense, so an inbuilt protection against the sun isn't necessary. When *Homo sapiens* arriving from Africa had sex with their northern neighbours, they passed these features on to their offspring permanently. Other characteristics from these dalliances are still active today, including a tendency towards depression and an addiction to tobacco products.[3] It was a two-way street, however, and our genes also entered Neanderthals, something scientists had for a long time rejected as impossible. In short, about 100,000 years ago, modern humans and their now-extinct cousins met and became intimate – so intimate that traces of these romantic rendezvous have been found in Neanderthal bones discovered in the Altai Mountains.[4]

All this research into Neanderthals is telling. We only attribute to this species of human characteristics that according to the current state of research are beyond dispute. Wouldn't it be more honest to say that although we know some things for certain, there are others about which we don't (yet) know enough? It seems suspiciously as though we can't bring ourselves to acknowledge that there are other beings as intelligent as we are, and nothing will be allowed to upset this belief. Not because it is forbidden, but because our instincts struggle against it with a resounding 'Never!'

Nature knows of only two paths for the future of every species: adapt or die out. And although these adaptations can include changes in intellectual capacity – which is what we're talking about here – we need to be clear: evolution means adapting to change, not necessarily development in the sense of improvements of the brain and its size.

Indeed, US researchers suspect that there are definite disadvantages to our powerful brain. They compared the self-destructive programming of human cells with a similar programme that works in great apes, specifically chimpanzees. This programme destroys and dismantles old and defective cells. Their comparison showed that the clean-up mechanism is a lot more effective in chimpanzees than it is in people, and the researchers believe that the reduced rate at which cells are broken down in humans allows for larger brain growth and a higher rate of connections between cells. But this improvement in intelligence probably comes at

a high price, because the chimps' self-cleansing mechanism also gets rids of cancer cells.[5] Whereas chimpanzees
hardly ever get cancer, in humans this disease is one of
the top causes of death. Are we paying the price for our
intellectual capacities? If our current level of intelligence
is not suited to the survival of humankind, it must either
be increased or lowered. The latter is probably unac-
ceptable, as we can't reconcile it with our ideas about
self-worth.

But if you leave aside the amazing things our large
brains allow us to do, we might ask whether the amount
of intellectual ability that we have today is really necessary for our personal quality of life. What is really
important? There's happiness, love and security, of
course, alongside such pleasures as delicious food and a
comfortable home. Do you notice something? All these
things have to do with feelings and instincts, and not
with intellectual achievements. People living in the year
50,000 CE will be able to live fulfilling lives independent
of the volume of their brains, as long as they can adapt
to constant change in their environment. And they will
have to – for no one can escape the network of nature.

16

The Old Clock

_____

NATURE IS CONSIDERABLY MORE COMPLICATED than the finely calibrated movement of a clock, but I would still like to return to the image I introduced at the opening of the book. We've now seen numerous examples of what happens if we carelessly remove one little cog. Just as in the internal mechanism of the timepiece, this loss triggers a chain reaction that changes the whole system.

But what if the clock stops working altogether, and we want to get it going again? We know in certain circumstances nature can heal itself, but we also know this takes time. Where natural processes take hundreds or thousands of years, could humans step in to speed things up? And since we like to enjoy the fruits of our labours – often that's what's really going on – we want to experience such improvements for ourselves. Why abstain from using vehicles that run on fossil fuels or avoid plastic when only our great-great-grandchildren will see the results of our efforts? And so we step in, determined to achieve positive change as quickly as possible. But when we take it upon ourselves to repair the internal workings of the environment, a big problem arises: how do we know when it's broken?

A good example of one of these attempts to put things right involves the capercaillie, or wood grouse. This large chicken-like bird (depending on the sex, it can weigh up to 4 kilograms) lives in boreal coniferous forests – that's to say, it's at home in the spruce and pine forests of the North. There it eats insects, but mostly it consumes the leaves and berries of blueberries. My family and I came across little fruit bushes everywhere when we were out and about in the forest in Lapland. And when we went walking in the mountains we saw capercaillie everywhere, too. Of course, we were excited every time one of these birds crossed our path, even though in northern Scandinavia they're nothing special. There they're considered wild game, and often end up in the pot.

Central Europe is different: here the birds are strictly protected. There are only a few areas that provide suitable habitats for capercaillie, because sufficiently large natural expanses of coniferous forest carpeted with blueberry bushes are found only in alpine regions. High up in the Alps, as in Scandinavia, winters are long and harsh, making it too cold for deciduous trees. So a handful of capercaillie live here just below the tree line. Of course, tiny scattered populations of any species are inherently unstable. If just a few of them die, the local population will not survive.

In the Middle Ages, the situation in central Europe was much better for these birds. Forest clearance created half-open landscapes where blueberry bushes grew in abundance. Today, you can find them in many planted coniferous forests, particularly in plantations of pines. As they're shaded by trees, the bushes rarely have any berries on them, but they are a reminder of earlier times when the practice of felling trees created the clearings where they love to grow.

This human activity suited the capercaillie too. They extended their range southwards and settled in habitats where they hadn't occurred originally. But with the arrival of modern forestry methods forests changed again. Pastures and agricultural land were reforested, and ravaged forests recovered and filled in. Deciduous trees returned full of vim and vigour to replace some of the gloomy coniferous plantations, and below their leafy branches it was much darker than underneath pines. Things were not looking good for blueberries and other bushes – nor, for that matter, for red wood ants, which no longer had access to the discarded needles they need to build their nests or the shafts of sunlight they require to warm them up enough to go about their business.

The death knell for the capercaillie and the blueberry was the rebirth of beech forests, the native vegetation of Germany. Is that a bad thing? I don't think so. All it means is that those species are being pushed back to the places they came from, and, in return, the rare inhabitants of German beech forests are getting their original habitats back.

You could say that everything is slowly recalibrating. Indeed you could. But now government and conservation organisations are getting involved. And we're back to the immense clockwork of nature. Is it really broken? Is there something that needs to be repaired? Unfortunately, this question is not even asked, at least not when it comes to the big picture. Instead, the capercaillie has been declared particularly worthy of conservation in the Black Forest, which was originally ancient deciduous forestland. With great fanfare, clearings have been made, and here and there the forest has even been burned to create open spaces for blueberries to grow. That Germany's native forest dwellers are now suffering – ground beetles, for example, which love the dark – is conveniently ignored.

It's a similar story for the capercaillie's smaller relative, the hazel grouse. Even finding its feathers near a construction site means work has to be halted immediately until the situation can be fully investigated. Hazel grouse are in danger of disappearing in Germany. Yet the Eifel originally contained nothing but ancient deciduous forests. Which means that the small hazel grouse would never have managed to make a life here if it hadn't been for people settling and clearing land, and creating large juniper heaths with their herds of cattle. In these new, lightly treed habitats – similar to forests in northern Sweden – hazel grouse felt quite at home. Unfortunately for them, however, the forests here are now recovering and shading out the juniper heathlands.

We can now see many different threads coming together. Conservationists who desperately want to help the grouse are pleading for the Eifel to be designated a protected habitat. They argue for more thinning of trees, which would mean more light reaching the ground, encouraging more bushes to grow, and so the basic food for this grouse would recover. Forest agencies are offering to step in and help. Isn't it time to revive coppicing? Coppicing is an old way of managing forests that arose hundreds of years ago out of pure necessity. When timber was becoming increasingly scarce, because it was being used so heavily as fuel and building material, people were no longer willing to give trees time to grow old. Oaks and beeches were being cut down at the tender age of 20 to 40 (instead of 160 or 200), because people just couldn't wait any longer. Acres of forest were razed. New shoots grew from the stumps, and that spindly growth was harvested just a few decades later.

So much of the forest was plundered that it began to look like a carpet full of holes. Hazel grouse love these conditions – and who can blame them? However, more sensible ideas about forestry then prevailed, and strict laws forbade coppicing. At least, they did until the modern need for timber, fired by the boom in bioenergy, set in. And so it was that the new clear-felling that came to be celebrated as reinvigorating a historical forest practice also helped the small hazel grouse.[1]

Romantic notions of the good old days and conservation combined? Not really. Now, as then, this is nothing but brutal clear-felling using automated harvesting machines that weigh many tonnes. This is no way to establish a proper forest, and whether the hazel grouse at the heart of this procedure really enjoys the newly cleared habitats remains to be seen. Meanwhile, things are not looking good for forest species that really belong here, such as the black woodpecker and that mealworm beetle we met earlier.

Retaining open meadows is another example. Meadows provide the habitat for a large number of grasses and non-woody plants. In summer they are full of colourful blooms and a-flutter with gaily painted butterflies. Such splendour attracts many different species of bird to settle here in great numbers. As agriculture becomes more intensive, this diversity is threatened. Thanks to the rise in the price of maize in response to the explosion in demand for raw materials for the biogas industry, every scrap of spare land is being ploughed under and converted into an agricultural desert devoted to a single crop. And even where the meadow idyll does still exist, the forest is preparing to reclaim the last isolated valleys and the wetlands along the banks of streams.

Grassy landscapes, then, are under siege. But instead of pointing the finger at agriculture, we are pitting grass against tree, which means that in order to preserve species that love meadows, it is forests and not farmed fields that have to give way. On the whole the methods used to combat the forests look thoroughly benign – reintroducing the Heck cattle I mentioned earlier, for example. These are supposedly back-crosses to aurochs, our original wild cattle, which once grazed the moist meadows alongside rivers and streams. Unfortunately, it isn't possible to breed these extinct species back to life, even if Heck cattle do bear a passing resemblance to ancient aurochs.

In the final reckoning, Heck cattle are nothing but domesticated cattle designed to look like their ancient ancestors. Allow these cattle to graze alongside streams and all looks right with the world. In reality, though, what this mode of farming does is reinforce a widespread misconception. Grass plains do not belong to the natural ecosystem in Germany, where there was once only forest punctuated by the occasional mountain range or swamp. The many colourful plants with their butterflies arrived on the coat-tails of human activity, and only became established when our ancestors cut down the trees.

There's a simple reason why these treeless land-scapes delight us so much. We are, from a biological perspective, animals of the plains, and we feel secure in landscapes with extensive views where we can move around easily. Do you remember that megaherbivore theory? It's brought in here too in the confusion between conservation and aesthetics to push the pendulum in favour of the latter. Yet if we were just to leave nature alone, then wetland forests would naturally regenerate along our streams and rivers. They don't support colourful plants and butterflies, but they do provide an important habitat for tens of thousands of other species. Think of the tree sap hoverfly. Until recently no one even knew it existed. If Heck cattle had created grassy landscapes by preventing the growth of moisture-loving trees, then this fly would have disappeared and we would have been none the wiser. In truth we don't understand the clockwork of nature, and as long as we don't, we shouldn't try to fix it.

Let me make myself clear. I don't object in principle to making a special effort to help individual species, even if the species is here as a result of human interference, like the hazel grouse and the capercaillie. If the species arrived in Germany in historical times and is now threatened globally with extinction, then – but only then – we should go out of our way to help it, even if it means upsetting parts of the native forest ecosystem. If there is no global threat, however, then any interference in the complex web of nature ought to be out of the question.

The red kite is a case in point. This raptor with its imposing wingspan of up to 180 centimetres is a perfect example of a species that has benefited from cultural changes in the landscape, and would certainly have been rare in the original ancient forests of central Europe. It needs open landscapes so it can glide through the air to hunt for small mammals, birds and insects. People's desire to clear forests suited these birds just fine. It created a plains environment with magnificent hunting possibilities.

Out in the fields every summer you can see how adaptable the red kite is. The moment a farmer starts cutting hay, there will often be a red kite following behind, looking for mice or fawns that have been minced or run over. Most of the global population of about 25,000 to 30,000 individuals lives in Germany. In most other places, red kites have declined drastically. If we now went back exclusively to our native vegetation, it would all be over for the majority of these birds. They have found a second home with us, and their populations are relatively healthy and should therefore be encouraged so that they remain so in the future. The best way to do that is by not only protecting a landscape of small farms and small fields but also by preserving the trees where they nest, by establishing buffer zones where there is no active forestry.

Just to remind you: we are talking here of intentional interference with natural processes. We interfere unintentionally a great deal everywhere all the time, and right now I'd like to restrict my examples to the open countryside. In most places in Germany, we have driven out the ancestral plants (trees) and replaced them with grains, potatoes and vegetables. What all cultivated species have in common is that they are not native to the area. Even in what remains of the forest, most of it is made up of non-native species. Wouldn't it be nice if, at least in protected areas, we allowed nature to take the helm?

If you think that goes without saying, just take a look at the paperwork produced for conservation areas and national parks. You'll find maintenance and development plans that are far too eager to put sawmills, chainsaws and heavy machinery to work. In the end, such plans for saving as many native species of trees as possible are neither aesthetically pleasing nor ecologically beneficial. We've already seen that most attempts at fixing things come to nothing, so why not simply trust mechanisms which are millions of years old to carry on functioning without us?

Amid all the awful news about the worldwide destruction of forests, there are some increasing signs of hope. More and more people want to protect existing forests and plant new ones. Nevertheless, this desire to start over raises a question: can we ever recreate these multifaceted ecosystems? The Brazilian rainforest offers some grounds for optimism. It is thought to be particularly vulnerable to changes wrought by civilisation, because the soils on which it depends are so old. I mean 'old' here in terms of geological eras during which these soils have barely changed at all. This is partly because since the Tertiary, which ended more than 2.6 million years ago, no new mountains have been formed, which means there has been barely any erosion or creation of new soil by weathering rocky slopes. This quiescent geological period is evident deep into the underground soil layers, and by deep I mean down to an impressive depth of 30 metres.

In most of the forest I manage, you can't dig down further than about 60 centimetres before you hit an underground layer of gravel, and even the upper layers of soil contain a lot of small stones. In many tropical soils in the Amazon, on the other hand, all the rock has been ground down into tiny particles. That might sound like nutrient-rich soil to you, but it's the exact opposite. Having been rained on for hundreds of thousands of years, the soils have lost most of their nutrients – water seeping through the soil has washed them down far deeper than plant roots can reach.

The wide range of species around today, as well as the exuberant growth of forests in these latitudes, seems to contradict this fact, but the fecundity of these forests is possible only because nutrients are held hostage in the rainforest ecosystem as an army of insects, fungi and bacteria recycles everything that dies by eating, digesting and excreting it. Every trunk that rots, every leaf that gets eaten by insects and is excreted as humus, releases the minerals it contains, which are greedily taken up by the trees' roots and converted into living tissue again. If you cut down the trees, this cycle of life is rudely interrupted.

Clearing by fire leaves a great deal of ash. Ash is simply nutrients in concentrated form, which are now exposed to heavy rains without any form of protection, meaning they are flushed away in rivers, disappearing forever from the land. This is why such clearances benefit agriculture for a short time only – until, that is, the brief boost of fertility from the ash fizzles out. The ravaged soils that remain are barely capable of supporting new trees, and any trees that planted here struggle to survive – if they survive at all. True tropical diversity with millions of species depends on the return of all the fungi, insects and vertebrates, and they require such special conditions that their return is unlikely. Or is it?

Let's go back to recovery's ground zero. The forest is gone and the soils are exhausted. How can there ever be any prospect of recovery if nutrients have disappeared deep underground never to be seen again, or been washed by the rain into the nearest river? After all, there's no natural mechanism to pump them back to the surface or float them back from the distant ocean. Yet the situation is not hopeless, and the tortured land doesn't necessarily have to become a desert.

As far as a minerals crisis is concerned, the Sahara could be seen as a first responder. Dust storms blow from the desert into the air an enormous quantity of tiny particles of dirt, which are then carried way up high on the wind from Africa to South America. There the dusty cargo is washed down by regular heavy rain to fertilise the ground. Nearly 30 million tonnes of desert dirt arrive this way every year, including about 22,000 tonnes of phosphorus, which is a potent plant fertiliser.

Scientists from the Earth System Science Interdisciplinary Center (ESSIC) at the University of Maryland[2] analysed seven years' worth of satellite images to estimate the amount of dust as accurately as they could. The figures varied enormously, but they strongly suspected that the constant arrival of airborne fertiliser is offsetting the nutrients washed away into the ground. That's the case for intact forests only, though, and when forests are felled there is a marked increase in the rate at which minerals are lost. Damn. There doesn't seem to be any way out of this mess. Is the situation really hopeless? No, as we can see if we take a closer look at that clear-felling in the Amazon. For when large expanses of forest are cleared away here, the remains of settlements appear – human settlements.

In the state of Acre in Brazil, a research team led by Jennifer Watling, now at the University of São Paulo, found 450 geoglyphs. These are earthworks laid out in geometric patterns, and in Acre they consist of a series of trenches and berms distributed across 13,000 square kilometres. People must have cleared the forest to build them, but the sites reveal that the original inhabitants proceeded with great caution. The researchers found no evidence of large-scale clearance. Instead, they discovered a system of forest management spanning thousands of years. But wait a minute. Discovered? How is it possible to calculate the size of clearings made in the forest thousands of years ago?

The key here turns out to be microscopic particles of silica called phytoliths, which are found in some plants. These particles vary from plant to plant, which is helpful, but what is even more important is that, unlike organic substances, which decompose quickly, these crystals last practically forever. And so, it is possible to build up a picture of what the vegetation looked like based on the frequency of different phytoliths.

Jennifer Watling and her team discovered that over the course of the 4,000 years the indigenous peoples were changing the forest in Acre, grass – a plant typically found in open spaces – never made up more than 20 per cent of the vegetation, and the combination of trees was significantly altered. Around the human-made constructions, the number of palm trees, important sources of both food and building materials, increased dramatically. Even today, more than 600 years after the settlements were abandoned, a noticeable number of palm trees remains in areas close to the geoglyphs.

The researchers' findings are encouraging. First, here we have an example of agroforestry – a mixture of agriculture and silviculture in the same area – that clearly functioned for a very long time without having any great impact on the environment. What worked then should also work now, and it points to how we could preserve as much forest as possible without excluding people. Second, after 600 years, the forest has regenerated so well that before this discovery, scientists assumed this was virgin forest untouched by human hand. So it seems we should put more trust in forest ecosystems and no longer use the word 'irretrievable' when describing how they have changed. And third, there is a message here about climate that really caught my attention.

The indigenous forest settlers carried out their system of land management over enormous areas, and as soon as they disappeared, the forest everywhere recovered on a similarly large scale. The small areas devoted to agriculture were quickly overgrown by trees, the density of the forest increased, and a great deal of carbon was stored in the mighty trees. In fact, so much carbon was absorbed all at once that the research team thinks it's possible that this was what triggered the Little Ice Age, and not the erupting volcanoes I mentioned earlier.[3] From the fifteenth century into the nineteenth century, temperatures fell, and failed harvests and famine went hand in hand with cold, rainy summers and long, freezing winters. Was all of this caused by the recovery of the Amazonian rainforest?

Of course, no one wants to return to times of famine, but our problem today isn't cold but increasingly warm temperatures. The positive message from all this is that not only can we win back the original forests, but that it could also steer the climate in the right direction. And to achieve this we don't even need to do anything. Quite the opposite, in fact. We need to leave things alone – on as large a scale as possible.

Epilogue

I LOVE TELLING STORIES. I also love playing the ukulele, even though I've been practising for years now and still can't play it very well. It's a bit different with story-telling, and that's because of the feedback I've received from the public (perhaps even from you). I remember the first time I appeared on television, back in 1998. In those days, I offered a survival course deep in the forest, where participants had to get through the weekend equipped with nothing but a sleeping bag, a cup and a knife. As you can imagine, this was a dream subject for the media (under the headline 'A Forester Who Eats Worms!'), and a camera crew from the local TV station came to the forest to interview one of the groups – and me, of course.

I thought I acquitted myself rather well, and later I sat down proudly with my family to watch the clip on the local news. But far from being impressed, they were soon pointing out the many 'er's that punctuated every sentence. 'There's another one, Dad!' my children called out every couple of seconds with increasing delight. My own pleasure, however, diminished with every remark they made, and by the time the programme finished I was in a pretty bad mood. With the next interview, therefore, I took pains to avoid saying 'er', and gradually I managed to garner a little praise for my appearances.

Something similar happened over the many guided tours I gave round the forest I manage, when I talked about things like ecological forest management or took people through our forest cemetery, the Final Forest. No one corrected me or remarked on my verbal short-comings, but there were always questions afterwards. I soon noticed I was using too many technical terms, and giving an impersonal, boring recitation of facts about something close to my heart: the wonderful forest ecosystem and the threat it was under. From the moment I started speaking the response was subtle but still painful. As soon as the first eyelids began to droop, I knew I was being too dry. So over the years, an emotional undertone crept in that was much more in tune with my own convictions. In other words, I relaxed, and let my heart do the talking instead of my brain.

Time and again after the guided tours were over, participants asked where they could read more on the things I'd been explaining. All I could do was shrug apologetically. At some point, my wife pressured me into putting at least a few pages together to give us something I could hand out to people interested in learning more. Back then, I had not the slightest interest in writing something like this. A friend suggested coming along on one of my tours with a recording device and turning the results into a book: I didn't like that idea much, either.

Then one day, on a holiday in Lapland, I sat down outside the camper van armed with pad and pencil and began to set down on paper the things I talked about on my tours. If after a year no publishing house had shown any interest in my book, I resolved, then I would cross writing off my to-do list once and for all. I hadn't expected things to turn out the way they did. A small press published my first book, *Wald ohne Hüter* (Forest without Guardians), and that, I thought, would be that. But as time went by, more books flowed from my pen, and I began to thoroughly enjoy myself.

Unfortunately, my work did not spark professional discussions among foresters about the way we treat forests. In hindsight, I can see that from the lobbyists' point of view it's best not to discuss such controversial issues publicly. But recently, when my book *The Hidden Life of Trees* came out, criticism from professional forestry circles came to a head. It was clearly in response to the pressure from the many people who'd read the book and wanted to know why we needed to set about our forests with all that heavy machinery. Instead of debating my ideas, however, most critics from the world of forestry took a different approach. The language I used was too emotional, they said. I described trees and animals in such a way as to make them seem human, and that wasn't scientifically correct.

But can a language stripped of emotion even be called a human language? Don't we function largely in response to our emotions? Are descriptions of nature only reliable when all processes are presented in biochemical terms and dissected so forensically that you get the impression plants and animals are fully automatic, genetically programmed biological machines? Of course, it would be possible to describe all our own feelings and activities in similar fashion, but that would fail to evoke what's really going on inside us and what enriches our lives. To me it's more important to state the facts in such a way that people can understand them emotionally. And then I can take them on a full sensory tour of nature to convey the joy our fellow creatures and their secrets can bring us.

Acknowledgements

THE NETWORK OF NATURE is too diverse to ever fit between the covers of a book, which means I was forced to choose particularly impressive examples and connect them so that readers could see the big picture. In this my wife, Miriam, was a great help. She read the manuscript many times with a critical eye. She didn't hesitate to point out passages that needed work, and she helped me see ways to improve my explanations.

As always, my children, Carina and Tobias, were a source of inspiration. Interminable debates around the breakfast table or in front of the television (which became a sort of electronic campfire) sparked new ideas that demanded space in the book.

My colleagues Lidwina Hamacher and Kerstin Manheller held the fort at the Forest Academy in Hümmel. It was during the time-consuming period of establishing the academy that I was working on the manuscript. Both of them understood when I was up against a deadline and needed to keep writing, and simply took over the management I should have been doing themselves.

My Mysteries of Nature trilogy – *The Hidden Life of Trees*, *The Inner Life of Animals* and this book – would never have happened if my publisher hadn't believed from the outset that my message to the people who visited the forest I managed should reach a wider audience. My agent, Lars Schultze-Kossack, helped me through all the issues that cropped up along the way.

Heike Plauert of Ludwig Verlag, my publisher, made it easy for me by placing all her trust in me and letting me get on with my writing. That suited me very well, as I was working on different sections of the book at the same time, a process that takes some getting used to. My editor, Angelika Lieke, tactfully helped me polish the manuscript.

Beatrice Braken-Gülke in the publicity department managed the media requests to give me room to breathe, though I would have been more than happy to answer all of their questions.

Last but not least, I would like to say a big thank-you to Jane Billinghurst. I am very pleased that her translations capture not only the meaning of the words I write but also the tone I wish to convey.

There are so many more people who are part of this process that unfortunately I cannot mention them all. From the printers to the distributors to the booksellers – every one has done their best to make sure you get to hold this book in your hands. And I want to thank you, dear reader, most sincerely for choosing this book from the many good books out there and joining me on this journey through nature.

Notes

1. Of Wolves, Bears and Fish

1. MyYellowstonePark.com, 'Wolf Reintroduction Changes Ecosystem', 21 June 2001, www.yellowstonepark.com/things-to-do/wolf-reintroduction-changes-ecosystem.
2. William J. Ripple et al., 'Trophic Cascades from Wolves to Grizzly Bears in Yellowstone', *Journal of Animal Ecology* 83, no. 1 (2014): 223–33, doi.org/10.1111/1365–2656.12123.
3. 'Der Lübtheener Wolf wurde gezielt erschossen'(The Lübtheener wolf was shot on purpose), NABU (Naturbundschutz Deutschland), December 2016, www.nabu.de/news/2016/12/21719.html.
4. M. Holzapfel et al., 'Die Nahrungsökologie des Wolfes in Deutschland von 2001 bis 2012' (The food ecology of the wolf in Germany from 2001 to 2012), www.wolfsregion-lausitz.de/index.php/nahrungszusammensetzung.
5. 'Wie viel Naturschutz verträgt unser Land?' (How much conservation can our country handle?), statement by Olaf Tschimpke, president of NABU (Naturbundschutz Deutschland), on the television programme *Hart aber fair* (Tough but fair), 23 January 2017, ARD (German Public Broadcasting), www.nabu.de/news/2017/01/21855.html.
6. A.D. Middleton et al., 'Grizzly Bear Predation Links the Loss of Native Trout to the Demography of Migratory Elk in Yellowstone', *Proceedings of the Royal Society B: Biological Sciences*

280, no. 1762 (15 May 2013): 2013.0870, doi.org/10.1098/rspb.2013.0870.

2. Salmon in the Trees

1. S.M. Gende, T.P. Quinn et al., 'Magnitude and Fate of Salmon-Derived Nutrients and Energy in a Coastal Stream Ecosystem', *Journal of Freshwater Ecology*, 19, no. 1: 149. doi.org/10.1080/02705060.2004.9664522.

2. J. Robbins, 'Why Trees Matter', *New York Times*, 11 April 2012, www.nytimes.com/2012/04/12/opinion/why-trees-matter.html.

3. T. Reimchen and M. Hocking, 'Salmon-Derived Nitrogen in Terrestrial Invertebrates from Coniferous Forests of the Pacific Northwest', *BMC Ecology* 2 (19 March 2002): 4, doi.org/10.1186/1472-6785-2-4.

4. C. Wolter, 'Nicht mehr als dreimal in der Woche Lachs' (Salmon no more than three times a week), *Nationalpark-Jahrbuch Unteres Odertal* 4: 118–26, www.nationalpark-unteres-odertal.de/de/publikationen/nicht-mehr-als-dreimal-der-woche-lachs.

5. 'Quatsch anfangen' (Talking nonsense), *Der Spiegel*, vol. 38 (1988).

6. ARGE Ahr, 'Bejagung des Kormorans' (Hunting cormorants), 25 October 2016, www.arge-ahr.de/tag/kormoran.

7. Anne Helmenstine, 'Elements in the Human Body and What They Do', *Science Notes*, 20 May 2015, sciencenotes.org/elements-in-the-human-body-and-what-they-do/.

8. A. Oita et al., 'Substantial Nitrogen Pollution Embedded in International Trade', *Nature Geoscience* 9 (25 January 2016): 111–15, doi.org/10.1038/NGEO2635.

9. Anne Post, 'Why Fish Need Trees and Trees Need Fish', *Alaska Fish & Wildlife News*, November 2008, www.adfg.alaska.gov/index.cfm?adfg=wildlifenews.view_article&articles_id=407.

3. Creatures in Your Coffee

1. Axel Bojanowski, 'Forscher rätseln über seltsame Tiefenwesen' (Researchers puzzle over strange life-forms), *Der Spiegel* online,

13 December 2013, www.spiegel.de/wissenschaft/natur/
mikroben-ursprung-des-lebens-kilometer-unter-erde-moeg-
lich-a-938358-druck.html.

2.   Bundesministerium   für   Umwelt,   Naturschutz   und
     Reaktorsicherheit (BMU), *Grundwasser in Deutschland*
     (Groundwater in Germany), Referat Öffentlichkeitsarbeit,
     Berlin, August 2008, 7.

3.   Gianfranco Novarino et al., 'Protistan Communities in Aquifers:
     A Review', *FEMS Microbiology Reviews*, 20 (1997): 261–75,
     doi:10.1111/j.1574–6976.1997.tb00313.x,onlinelibrary.wiley.
     com/doi/10.1111/j.1574–6976.1997.tb00313.x/full.

4.   R. Sender, S. Fuchs, R. Milo, 'Revised Estimates for the
     Number of Human and Bacteria Cells in the Body', *PLoS
     Biol* 14, no. 8 (2016): e1002533, doi.org/10.1371/journal.
     pbio.1002533.

4. Why Deer Taste Bad to Trees

1.   B. Ochse et al., 'Salivary Cues: Simulated Roe Deer Browsing
     Induces Systemic Changes in Phytohormones and Defence
     Chemistry in Wild-Grown Maple and Beech Saplings',
     Functional Ecology, Volume 31, Issue 2 (8 August 2016): 340–49,
     doi.org/10.1111/1365–2435.12717.

5. Ants – Secret Sovereigns

1.   Christoph Drösser, 'Ein Haufen Ameisen' (A whole heap of
     ants), *Die* Zeit online, 20 March 2008, www.zeit.de/2008/13/
     Stimmts-Ameisen-und-Menschen.

2.   W. Jirikowski, 'Wichtige Helfer im Wald: hügelbauende
     Ameisen' (Important forest helpers: hill-building ants), *Der
     Fortschrittliche Landwirt*, Graz (14): 105–7.

3.   T. Oliver et al., 'Ant Semiochemicals Limit Apterous Aphid
     Dispersal', *Proceedings of the Royal Society B*, 274, no. 1629 (22
     December 2007): 3127–32, doi.org/10.1098/rspb.2007.1251.

4.   T. Mahdi and J.B. Whittaker, 'Do Birch Trees (*Betula pendula*)
     Grow Better if Foraged by Wood Ants?', *Journal of Animal
     Ecology*, 62, no. 1 (January 1993): 101–16, doi.org/10.2307/5486.

5.   J.B. Whittaker, 'Effects of Ants on Temperate Woodland Trees',
     in C.R. Huxley and D.F. Cutler, eds, *Ant-Plant Interactions* (New
     York: Oxford University Press, 1991), 67–79.

6. Is the Bark Beetle All Bad?

1.   Government of British Columbia, 'Mountain Pine Beetle
     Projections', www2.gov.bc.ca/gov/content/industry/forestry/
     managing-our-forest-resources/forest-health/forest-pests/
     bark-beetles/mountain-pine-beetle/mpb-projections.
2.   H. Rosner, 'The Bug That's Eating the Woods', *National
     Geographic*, April 2015, ngm.nationalgeographic.com/2015/04/
     pine-beetles/rosner-text.

7. The Funeral Feast

1.   X. Gu and R. Krawczynski, 'Tote Weidetiere – staatlich verhinderte
     Forderung der Biodiversität' (Dead grazing animals – federally
     impeded requirements for biodiversity), *Artenschutzreport*, no. 28
     (2012): 60–64.
2.   Ibid.
3.   'Die Rückkehr des Knochenfressers' (The return of the bone
     eater), spectrum.de, 24 September 2010, www.spektrum.de/
     news/die-rueckkehr-des-knochenfressers/1046860.
4.   Club300.de, 'Rarities Germany: Gänsegeier' (German rarities:
     griffon vultures), www.club300.de/alerts/index2.php?id=203.
5.   C. Westerhaus, 'Weibchen lassen Männchen während der
     Brutpflege abblitzen' (Females rebuff males when caring for
     their young), Deutschlandfunk, 23 March 2016, www.
     deutschlandfunk.de/totengraeber-kaefer-weibchen-lassen-
     maennchen-waehrend-der.676.de.html?dram:article_id=
     349257.

8. Bring Up the Lights!

1.   H.U. Schnitzler and O.W. Henson, 'Sonar Systems in
     Microchiroptera', in R.G. Bushnel et al., *Animal Sonar Systems*
     (New York: Plenum Press, 1980).

2.   Hannah M. Moir et al., 'Extremely High Frequency Sensitivity
     in a "Simple" Ear', *Biology Letters*, 9, no. 4 (8 May 2013), doi.
     org/10.1098/rsbl.2013.0241.

3.   'Wo tanzt das Glühwürmchen?' (Where does the glow-worm
     dance?), www.laternentanz.eu/Content/Informations/Living.
     aspx.

4.   David J. Merritt and Sakiko Aotani, 'Circadian Regulation of
     Bioluminescence in the Prey-Luring Glowworm, *Arachnocampa
     flava*', *Journal of Biological Rhythms*, 23, no. 4 (August 2008):
     319–29, doi.org/10.1177/0748730408320263.

5.   Wynne Parry, 'Fireflies' Unique Flashes Help Distinguish
     Species', LiveScience, 29 March 2012, www.livescience.
     com/19376-firefly-glow-signals.html.

6.   T. Eisner et al., 'Firefly "Femmes Fatales" Acquire Defensive
     Steroids (Lucibufagins) from Their Firefly Prey', *PNAS*,
     94, no. 18 (2 September 1997): 9723–28, doi.org/10.1073/
     pnas.94.18.9723.

9. Sabotaging the Production of Iberian Ham

1.   Michael Stang, 'Die eigenwilligen Flugrouten der Kraniche'
     (The self-selected migration routes of cranes), www.deutschland-
     funk.de/globales-kommunikationsnetz-bei-zugvoeglen-die.676.
     de.html?dram:article_id=321788.

2.   Gregor Rolshausen et al., 'Contemporary Evolution of
     Reproductive Isolation and Phenotypic Divergence in Sympatry
     along a Migratory Divide', *Current Biology*, 19, no. 24 (3
     December 2009): 2097–101, doi.org/10.1016/j.cub.2009.10.061.

3.   Katy Sewall, 'The Girl Who Gets Gifts from Birds', *BBC News
     Magazine* online, www.bbc.com/news/magazine-31604026.

10. How Earthworms Control Wild Boar

1.   W. Arnold et al., 'Nocturnal Hypometabolism as an Over-
     wintering Strategy of Red Deer (*Cervus elaphus*)', *American
     Journal of Physiology, Regulatory, Integrative, and Comparative
     Physiology*, 286, no. 1 (1 January 2004): R174–R181, doi.
     org/10.1152/ajpregu.00593.2002.

2.   Blick Acktuell, 'Hohe Rotwilddichte im Kesselinger Tal wird zu Problem' (The high density of red deer in the Kesslinger Valley is becoming a problem), www.blick-aktuell.de/Bad-Neuenahr/Hohe-Rotwilddickte-im-Kesselinger-Tal-wird-zu-Problem-27341.html.

3.   U. Dohle, 'Besser: Wie mästet Deutschland?' (Better: How does Germany fatten up?), *Ökojagd*, February 2009, 14–15.

4.   Forstbotanische Garten (forest-botanical garden), Georg-August-Universität-Göttingen, 'Stieleiche' (The common oak), www.uni-goettingen.de/de/blüten-samen-und-früchte/16692.html.

5.   N. Hahn, 'Raumnutzung und Ernährung von Schwarzwild' (Territorial use and feeding of wild boar), *LWF aktuell* 35, 32–34, www.waldwissen.net/wald/wild/management/lwf_raum_schwarzwild/index_DE.

6.   Axel Weiss, 'Sauenmast im Westerwald' (Feeding sows in Westerwald), www.swr.de/blog/umweltblog/2008/10/18/sauenmast-im-westerwald/.

7.   Karl-Maria ImBoden, 'Regenwurm' (Earthworm), www.regen-wurm.ch/de/leistungen.html.

8.   S. Blome and M. Beer, 'Afrikanische Schweinepest' (African swine pest), *Berichte aus der Forschung*, *FoRep* 2/2013, Friederich-Löffler-Institut, Insel Reims.

11. Fairy Tales, Myths and Species Diversity

1.   Bundesamt für Naturschutz (BfN) (Federal Agency for Nature Conservation), 'Artenschutz-Report 2015, Tiere und Pflanze in Deutschland' (Conservation report 2015, Animals and plants in Germany), Bonn, May 2015, 12.

2.   Federal Ministry of Food and Agriculture, 'The Forests in Germany: Selected Results of the Third National Forest Inventory', October 2014, translated January 2015, www.bmel.de/SharedDocs/Downloads/EN/Publications/ForestsInGermany-BWI.pdf?__blob=publicationFile.

3.   'Neue Tierart entdeckt' (New species discovered), Pressemitteilung des Helmholtz-Zentrums für Umweltforschung UFZ (press release from the Helmholtz Centre for Environmental Research – UFZ), 20 March 2015, www.ufz.de/index.php?de=35747.

4. E. Dressaire et al., 'Mushroom Spore Dispersal by Convectively Driven Winds', Cornell University Library, December 2015, arXiv 1512.07611v1 [physics.bio-ph].

5. C. Pietschmann, 'Pilzgespinst in Wurzelwerk' (Fungal threads in root systems), Max-Planck-Institut für molekulare Pflanzenphysiologie (Max Planck Institute for Molecular Plant Physiology), 21 December 2011, www.mpimp-golm.mpg. de/5630/news_publications_4741538.

6. Anne Casselman, 'Strange but True: The Largest Organism on Earth Is a Fungus', *Scientific American*, 4 October 2007, www.scientificamerican.com/article/strange-but-true-largest-organism-is-fungus/.

7. G. Möller, 'Struktur- und Substratbindung holzbewohnender Insekten, Schwerpunkt Coleoptera-Käfer' (Structure and substrate-specific binding of insects that live in wood, with a focus on Coleoptera), Dissertation zur Erlangung des akademischen Grades des Doktors der Naturwissenschaften (Dr.rer.nat.), eingereicht im Fachbereich der Biologie, Chemie, Pharmazie der Freien Universität Berlin (PhD dissertation in natural sciences, Faculty of Biology, Chemistry, Pharmacy, Free University Berlin), March 2009, 35–36.

12. What's Climate Got to Do with It?

1. K. Naudts et al., 'Europe's Forest Management Did Not Mitigate Climate Warming', *Science*, 351, no. 6273 (5 February 2016): 597–600, doi.org/10.1126/science.aad7270.

2. Michelle Hampson, 'Centuries of European Forest Management Have Not Cooled Climate', American Association for the Advancement of Science, 3 February 2016, www.aaas.org/ news/science-centuries-european-forest-management-have-not-cooled-climate.

3. Jason Kirby et al., 'Ion-Induced Nucleation of Pure Biogenic Particles', *Nature*, 533 (26 May 2016): 521–26, doi.org/10.1038/ nature17953.

4. R. Wengenmayr, 'Staub, an dem Wolken wachsen' (Dust that grows clouds), Mitteilung der Max-Planck-Gesellschaft, 22 February 2010, Max Planck Institute for Chemistry, Mainz.

5. M. Dobbertin and A. Giuggiolo, 'Baumwachstume und erhöhte Temperaturen' (Tree growth and rising temperatures), *Forum für Wissen* 206: 35–45.

6. Institute of Veterinary Public Health, University of Veterinary Medicine Vienna, World Maps of Köppen-Geiger Climate Classification, koeppen-geiger.vu-wien.ac.at/.

7. Karl Gartner and Markus Neumann, 'Wie schützen sich Waldbäume vor extremer Kälte?' (How do forest trees protect themselves against extreme cold?), waldwissen.net, 6 March 2012, www.waldwissen.net/wald/klima/wandel_co2/bfw_schrumpfen_baumstamm/index_DE.

8. G.H. Miller et al., 'Abrupt Onset of the Little Ice Age Triggered by Volcanism and Sustained by Sea-Ice/Ocean Feedbacks', *Geophysical Research Letters*, 39, no. 2 (January 2012): L02708, doi.org/10.1029/2011GL050168.

13. It Doesn't Get Any Hotter Than This

1. D. Kraus, F. Krumm and A. Held, 'Feuer als Stöfaktor in Wäldern' (Fire as a forest disruptor), *FVA-einblick*, 3/2013, 21–23, www.waldwissen.net/waldwirtschaft/schaden/brand/fva_waldbrand_artenvielfalt/index_DE.

2. F. Berna et al., 'Microstratigraphic Evidence of in Situ Fire in the Acheulean Strata of Wonderwerk Cave, Northern Cape Province, South Africa', *PNAS*, 109, no. 20: E1215–E1220, doi.org/10.1073/pnas.1117620109.

3. P. Bethge, 'Ich koche, also bin ich' (I cook, therefore I am), *Der Spiegel*, 52, 2007, 126–29.

4. Peter Hirschberger, *Wälder in Flammen: Ursachen und Folgen der weltweiten Waldbränder* (Forests in flame), WWF Deutschland, Berlin, July 2011, 33.

14. Our Role in Nature

1. Monash University, News and Events, 'Humans Caused Australia's Megafaunal Extinction', 20 January 2017, www.monash.edu/news/articles/humans-caused-australias-megafaunal-extinction.

2. Thomas Litt, 'Waldland Mitteleuropa: Die Megaherbivorentheorie aus paläobotanischer Sicht' (Central Europe as forest: The mega-herbivore theory from a paleobotanical perspective), in Tagungsband, 'Grosstiere als Landschaftsgestalter – Wunsch oder Wirklichkeit?' (conference transcript: Large animals as shapers of the landscape – fact or fiction?), Freising, August 2000, Bayerisches Staatsministerium für Ernährung, Landwirtschaft und Forsten (Bavarian Ministry of Food, Agriculture and Forestry), no. 27: 50–57, www.gbv.de/dms/tib-ub-hannover/320198219.pdf.

3. Hubert Weiger, 'Chancen und Risiken der Megaherbivorentheorie' (Prospects and risks of the megaherbivore theory), in Tagungsband 'Grosstiere als Landschaftsgestalter – Wunsch oder Wirklichkeit?' (conference transcript: Large animals as shapers of the landscape – fact or fiction?), Freising, August 2000, Bayerisches Staatsministerium für Ernährung, Landwirtschaft und Forsten (Bavarian Ministry of Food, Agriculture and Forestry), no. 27: 3–5, www.gbv.de/dms/tib-ub-hannover/320198219.pdf.

4. BUND, Friends of the Earth Germany, 'Ein Rettungsnetz für die Wildkatze' (A safety net for wild cats), www.bund.net/themen_und_projekte/rettungsnetz_wildkatze/.

5. J.R. Moore et al., 'Anthropogenic Sources Stimulate Resonance of a Natural Rock Bridge', Geophysical Research Letters, 43, no. 18 (28 September 2016): 9669–76, doi.org/10.1002/2016 GL070088.

15. The Stranger in Our Genes

1. F. Ramirez Rozzi et al., 'Cutmarked Human Remains Bearing Neandertal Features and Modern Human Remains Associated with the Aurignacian at Les Rois', Journal of Anthropological Sciences, 87 (2009): 153–85.

2. C. Simonti et al., 'The Phenotypic Legacy of Admixture between Modern Humans and Neanderthals', Science, 351, no. 6274 (2 February 2016): 737–41, doi.org/10.1126/science. aad2149.

3. Ibid.

4.   M. Kuhlwilm et al., 'Ancient Gene Flow from Early Modern Humans into Eastern Neanderthals', *Nature*, 530 (25 February 2016): 429–33, doi.org/10.1038/nature16544.

5.   G. Arora et al., 'Did Natural Selection for Increased Cognitive Ability in Humans Lead to an Elevated Risk of Cancer?', *Medical Hypotheses*, 73, no. 3 (September 2009): 453–56.

16. The Old Clock

1.   Example of how coppicing is portrayed in the promotional literature of a federal forest commission: Landesforsten Rheinland-Pfalz (State Forests of Rhineland-Palatinate), 'Niederwald in Rheinhessen – ein neues Projekt' (Coppicing in Rheinhessen – a new project), www.wald-rlp.de/en/forstamt-rheinhessen/wald/niederwaldprojekt.

2.   H. Yu et al., 'The Fertilising Role of African Dust in the Amazon Rainforest: A First Multiyear Assessment Based on Data from Cloud-Aerosol Lidar and Infrared Pathfinder Satellite Observations', *Geophysical Research Letters*, 42 (18 March 2015): 1984–91, doi.org/10.1002/2015GL063040.

3.   J. Watling et al., 'Impact of Pre-Columbian Geoglyph Builders on Amazonian Forests', *PNAS*, 114, no. 8 (2017): 1868–73, doi.org/10.1073/pnas.1614359114.